Learning from Data Streams

João Gama • Mohamed Medhat Gaber

Learning from Data Streams

Processing Techniques in Sensor Networks

With 73 Figures

Editors

João Gama
Laboratory of Artificial Intelligence and Decision Support, INESC-Porto LA
and Faculty of Economics
University of Porto
Rua de Ceuta, 118, 6º
4050-190 Porto, Portugal
jgama@liacc.up.pt

Mohamed Medhat Gaber
Tasmanian ICT Centre
GPO Box 1538
Hobart, TAS 7001
Australia
Mohamed.Gaber@csiro.au

Library of Congress Control Number: 2007931339

ACM Classification: H.3, C.2, C.3, I.2

ISBN 978-3-540-73678-3 Springer Berlin Heidelberg New York

Springer is a part of Springer Science+Business Media
springer.com

Typesetting: by editors
Production: VTEX Ltd., Lithuania
Cover Design: KünkelLopka, Heidelberg

Printed on acid-free paper 45/3180/VTEX 5 4 3 2 1 0

Preface

Data streams are everywhere these days. Look around and you will find sources of information that are continuously generating data. Our everyday life is now getting stuffed with devices that are emanating many data streams. Cell-phones, cars, security sensors, and televisions are just some examples. More are on their way. Smart houses of the future are likely to have many different types of sensors for sensing and adapting the domestic services to the changing needs of the inhabitants. For example, artificial motion, audio, and visual sensory systems are already being used to revolutionize the face of assisted at-home living for senior citizens. Cars are also increasingly becoming rich sources of data. The vehicle data bus and onboard GPS devices are tapped for extracting vehicle-health, fuel consumption, driver behavior, and location-dependent data. Data streams are common in many commercial and law-enforcement-related surveillance applications. Many of the public places that we often visit (e.g. airports, metros) and places where we work are monitored round the clock using various types of sensors. Perimeters of secured establishments and borders of countries are getting monitored using sensors. Wearable computers embedded in clothes and accessories for measuring our body parameters are increasingly getting popular among sport enthusiasts, patients, military, and researchers studying human computer interaction. In short, data streams are widely prevalent and in fact these days it is somewhat hard to find any source of information that does not generate continuous streams of data.

Extracting knowledge from multiple, possibly distributed data streams, is one of the most significant challenges that we face today. Conventional algorithms for analyzing static data sets that work based on many assumptions such as stationary distribution, ability to store all the data in memory at the same time, and centralized collection of all the data for subsequent analysis are questionable in this new domain—analyzing data streams in multi-sensory distributed environment. We need to understand the issues better and develop efficient data mining algorithms and systems for streams. Data mining deals with data analysis problems by paying attention to computing, communication, storage, and human-factors. Likewise, data stream mining technology must deal with those issues in the context of streams and distributed environments.

This book offers a high quality overview of the field of data stream mining for networks of sensors. The editors have made a commendable effort in putting together a nice collection of chapters that deal with data stream mining algorithms, systems, and applications. The first part of the book presents an overview of the field of data stream processing and sensor networks. The second part considers issues in data stream management. The third part discusses data stream mining algorithms. Finally, the last part presents some of the exciting applications. The structure of the book is balanced and I am really pleased that the editors have decided to put together a book at a time when the community really needs that. I congratulate the editors and the authors for publication of this book. I have enjoyed reading it and I hope you would do the same.

June 15, 2007 Baltimore, USA Hillol Kargupta

Contents

Part II Data Stream Management Techniques in Sensor Networks

Part III Mining Sensor Network Data Streams

Part IV Applications

Chapter 1
Introduction

João Gama and Mohamed Medhat Gaber

A cloud of nanoparticles—micro-robots—has escaped from the laboratory. This cloud is self-sustained and self-reproducing. It is intelligent and learns from experience. For all practical purposes, it is alive.
Prey, Michael Crichton, 2002

1.1 Preamble

Sensor networks consist of distributed autonomous devices that cooperatively monitor an environment. Each node in a sensor network is able to sense, process and act. Sensors are equipped with capacities to store information in memory, process information and communicate with neighbors. They have strong constraints on resources such as energy, memory, computational speed and bandwidth.

Typical applications of sensor networks include monitoring, tracking, and controlling. Some of the specific applications are habitat monitoring, object tracking, nuclear reactor controlling, fire detection, traffic monitoring etc. In a typical application, a wireless sensor network (WSN) is scattered in a region where it is meant to collect data through its sensor nodes. Sensor nodes can be imagined as small computers, equipped with basic capacities in terms of their interfaces and components. Sensors act in dynamic environments, under adversarial conditions.

While the technical problems associated with sensor networks is achieving a stable phase, very few works address the problem of analyzing and automatically un-

J. Gama
LIAAD, University of Porto, R. de Ceuta 118-6, 4050-190 Porto, Portugal
e-mail: jgama@fep.up.pt

M.M. Gaber
Tasmanian ICT Centre, CSIRO ICT Centre, GPO Box 1538, Hobart, TAS 7001, Australia
e-mail: Mohamed.Gaber@csiro.au

derstanding what is going on in the data produced. Learning from data produced by sensor networks poses several new problems: sensors are distributed; they produce a continuous flow of data, eventually at high speeds; they act in dynamic and time-changing environments; the number of sensors can be very large etc.

The input for machine learning and data mining algorithms is data. Sensor networks produce type-changing data streams. A data stream is a sequence of unbounded, real-time data records that are characterized by the very high data rate, which stresses our computational resources, and can be read only once by processing applications [13,8,1,9]. Research in data stream processing has brought efficient data mining [8,7] and management [1] techniques. Although this research has been driven by many application domains, the lack of killer applications has been an obstacle. With the evolution of sensing hardware, sensor networks have attracted the attention of data stream researchers as a platform for many significant applications.

Processing data streams generated from wireless sensor networks has raised new research challenges over the last few years [5,3,6]. Sensor nodes have limited power to send all of their measurements to a central high-performance computational facility to be processed. In-network data processing represents the acceptable mode of operation in sensor networks. Resource constraints of individual sensor nodes add new challenges to the problem. On the other hand, central processing is still a valid approach for some applications that do not require all the data to be processed but rather use data taken at longer intervals.

Applications of data stream processing in sensor networks include security, scientific, business and industry [4]. Owing to the significance of its applications, research in this area has attracted researchers from different areas. These include mainly networking [11], database [10,12,14] and data mining [3,2].

The focus of this book is to provide the reader an idea of the state-of-the-art in data stream management and mining in sensor networks. The rest of this chapter gives an overview of the book and provides a roadmap for the reader.

1.2 Book Overview

The book consists of four parts. Part I provides an overview of sensor networks. This part has three other chapters as well. Since the book is concerned with data stream processing in sensor networks, overviews about sensor networks, data stream processing in general, and data stream processing in sensor networks in particular are provided in this part as follows. In Chap. 2, Barros provides a background about sensor networks from technological and computational points of view. In Chap. 3, Gama and Rodrigues give an overview of data stream models, algorithmic methodologies and research issues and challenges. In Chap. 4, Gaber classifies data stream processing in sensor networks and identifies the research issues and challenges that face sensor networks with relation to data stream processing.

Part II of this book discusses data stream management techniques in sensor networks. In Chap. 5, Hammad, Ghanem, Aref, Elmagarmid and Mokbel present their data stream management system, Nile. The features of Nile have been extensively discussed. Although Nile is designed as a generic data stream management system, its features make it a strong tool for data stream management in sensor networks. In Chap. 6, Trigoni, Guitton and Skordylis discuss query processing in sensor networks. In Chap. 7, Shrivastava and Buragohain discuss issues on Aggregation and summarization of sensor data streams. Finally, in Chap. 8, Cardell-Oliver survey techniques to monitor data streams in sensor networks.

Part III of this book is concerned with data stream mining in sensor networks. In Chap. 9, Rodrigues and Gama discuss clustering techniques. In Chap. 10, Gama and Pederson discuss centralized and distributed predictive learning. Finally, in Chap. 11, Sun, Papadimitriouet, and Yu provide a detailed description with experiments of their method for using tensor analysis in sensor networks.

The book concludes with a look at applications. In Chap. 12, Ganguly, Omitaomu and Walker discuss security applications. Ganguly, Fang, Khan, and Omitaomu, in Chap. 13, present scientific applications in wireless sensor networks. In Chap. 14, Pederson illustrates the use of sensor networks in education using Lego mindstorms.

1.3 Roadmap

This book is organized to facilitate reading from different perspectives. Figure 1.1 shows the book roadmap. Readers interested in having an overview of the area with a sense of real-world applications are recommended to read Parts I and IV. On the other hand, data stream management researchers and practitioners can read Parts I and II. Finally, readers who are interested in data stream mining can read Parts I and III.

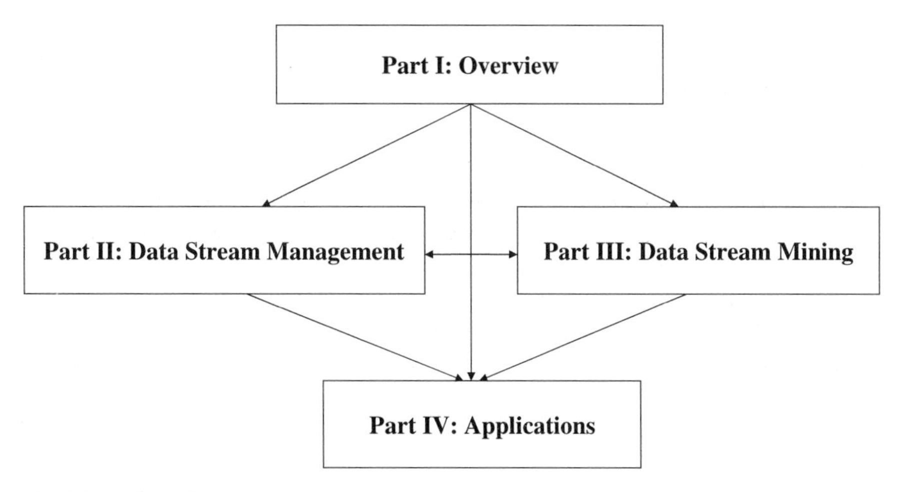

Fig. 1.1 Book roadmap

Generally, the book could be read in any order according to the reader's interest. Chapters are not dependent in the sense that any chapter should be read before the other. However, we strongly encourage the reader to start with the overview in order to have the required background to read the book.

1.4 Final Remarks

The future will witness large deployments of sensor networks. These networks of small sensing devices will change our lifestyle. With the advances in their computational power, such networks will play a number of roles, from smart houses and offices to roads and other transportation systems. They will be part of our daily activities. This view has started in science fiction novels, which tend to deal with extremes. In scientific fiction the concept of *swarm intelligence* is used in the context of groups of several small artificial artifacts that sense the environment, process local information and communicate with neighbors. Swarm intelligence is used to solve problems like destroying cancer cells, exploiting extra-solar planets, emulating social complex dynamics etc. Nowadays we are witnessing some of the realities. This book is not science fiction, it's more a small step towards the future.

References

[1] B. Babcock, S. Babu, M. Datar, R. Motwani, J. Widom, Models and issues in data stream systems. In: Proceedings of Principles of Database Systems (PODS'02), pp. 1–16, 2002.
[2] G. Boone, Reality mining: browsing reality with sensor networks. In: Sensors Online, vol. 21, 2004.
[3] V. Cantoni, L. Lombardi, P. Lombardi, Challenges for data mining in distributed sensor networks. ICPR (1) 1000–1007, 2006.
[4] D. Culler, D. Estrin, M. Srivastava, Overview of sensor networks. IEEE Computer, 37(8):41–49, 2004. Special Issue in Sensor Networks.
[5] E. Elnahrawy, Research directions in sensor data streams: solutions and challenges. DCIS Technical Report DCIS-TR-527, Rutgers University, 2003.
[6] J. Elson, D. Estrin, Sensor networks: a bridge to the physical world. In: Znati, Radhavendra, Sivalingam (Eds.) Chapter in Wireless Sensor Networks, pp. 3–20. Kluwer, Dordrecht, 2004.
[7] M.M. Gaber, S. Krishnaswamy, A. Zaslavsky, On-board mining of data streams in sensor networks. In: S. Badhyopadhyay, U. Maulik, L. Holder, D. Cook (Eds.), A book chapter in Advanced Methods of Knowledge Discovery from Complex Data, pp. 307–336. Springer, Berlin, 2005.
[8] M.M. Gaber, A. Zaslavsky, S. Krishnaswamy, Mining data streams: a review. ACM SIGMOD Record, 34(2):18–26, 2005.
[9] M. Garofalakis, J. Gehrke, R. Rastogi, Querying and mining data streams: you only get one look a tutorial. In: Proceedings of the 2002 ACM SIGMOD International Conference on Management of Data, June 03–06, Madison, Wisconsin, 2002.
[10] J. Gehrke, S. Madden, Query processing in sensor networks. IEEE Pervasive Computing, 3(1):46–55, 2004.

[11] B. Krishnamachari, Networking Wireless Sensors. Cambridge University Press, Cambridge 2006.
[12] S. Madden, M. Franklin, J. Hellerstein, W. Hong, TinyDB: an acqusitional query processing system for sensor networks. ACM TODS, 30(1):122–173, 2005.
[13] S. Muthukrishnan, Data streams: algorithms and applications. In: Proceedings of the Fourteenth Annual ACM-SIAM Symposium on Discrete Algorithms, 2003.
[14] B. Scholz, M.M. Gaber, T. Dawborn, R. Khoury, E. Tse, Efficient time triggered query processing in wireless sensor networks. In: Proceedings of the International Conference on Embedded Systems and Software ICESS-07. Springer, Berlin, 2007.

Part I
Overview

Chapter 2
Sensor Networks: An Overview

João Barros

2.1 Sensing and Communicating

Whether by telephone, on television or over the Internet, a substantial part of our daily exchange of information occurs in a virtual world we call cyberspace. In the past decades, by introducing data into these networks and acting upon the collected data we, the human users of these networks, have been the sole bridges between the physical world we live in and this virtual world we use to communicate. With the advent of tiny, low-cost devices capable of sensing the physical world and communicating over a wireless network, it becomes more and more evident that the status quo is about to change—sensor networks will soon close the gap between cyberspace and the real world.

There are two important reasons why this vision is gaining momentum: (1) its inherent potential to improve our lives. With sensor networks we can expand our environmental monitoring, increase the safety of our buildings, improve the precision of military operations, provide better health care, and give well-targeted rescue aid, among many other applications. (2) The endless possibilities such networks offer for multi-disciplinary research combining typical areas of electrical and computer engineering (sensor technology, integrated systems, signal processing, wireless communications), with classical computer science subjects (routing, data processing, database management, machine learning, data mining, artificial intelligence), and all potential application fields (medicine, biology, environmental sciences, agriculture, etc.).

In many ways, sensor networks are significantly different from classical wireless networks such as cellular communications systems and wireless local area networks:

J. Barros
Instituto de Telecomunicações, Department of Computer Science, Faculdade de Ciências da Universidade do Porto, Porto, Portugal
e-mail: barros@dcc.fc.up.pt

(a) The design of a sensor network is strongly driven by its particular application.
(b) Sensor nodes are highly constrained in terms of power consumption, computational complexity and production cost.
(c) Since the network is dense and the nodes share a common objective—to gather and convey information—cooperation can be used to enhance the network's efficiency.

These key features lead to very challenging research problems that are best illustrated with a practical example.

Example 1 (Sensor Webs for Precision Farming). Precision agriculture is about bringing the right amount of water, fertilizer and pesticides to the right plantation site at the right time [11]. Breaking with traditional methods that spread excessive quantities of chemicals uniformly over a field, this new paradigm guarantees a more efficient use of the farm's resources, a strong reduction of undesirable substances on the soil and in the ground water, and ultimately better crops at lower costs. Since fundamental parameters such as soil moisture and concentration of nutrients depend on the measuring spot and can vary quickly in time, precision farming requires constant sampling of the site characteristics on a fine spatial grid and a short-time scale—an ideal application for wireless sensor networks. With the aid of a sensor web, the control center can obtain constant updates of the soil conditions site by site, perform the necessary data fusion steps and adapt the flows of water, fertilizer and pesticide on the go according to the actual needs of the growing plants, as illustrated in Fig. 2.1.

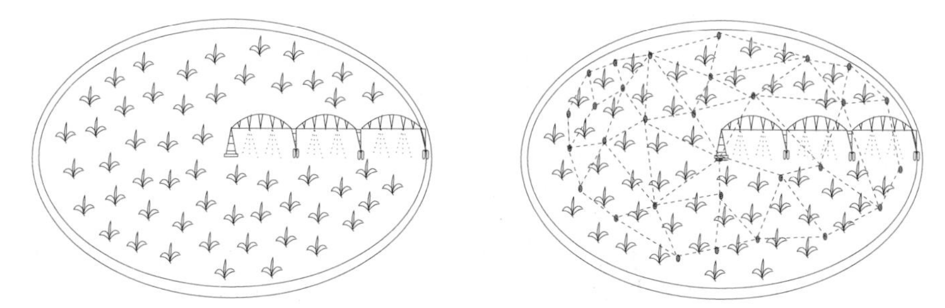

Fig. 2.1 A sketch of a center-pivot irrigation system. In the classical application (*left drawing*) the irrigation arm moves in circles and spreads water, fertilizer and pesticide uniformly over the crop area. To implement precision farming and increase the agricultural efficiency of the plantation, we can use a self-organizing sensor web that monitors the crop (*right drawing*) and sends the relevant data (e.g., soil humidity and nutrient status) to a fusion center over a wireless network. Based on the acquired information the fusion center can determine the right flow of water and chemicals for each sector of the crop

While the decision-making process falls within the scope of the agricultural engineer, the design of the supporting sensor web opens a myriad of challenging research problems for the system and software developers. Ultimately, as sensor networks evolve towards *sensor-actuator networks*, the convergence of computation, communication and control methodologies seems both inevitable and highly beneficial.

In the course of this introductory chapter, we will view the sensor network as a collection of transmitters that observe multiple sources of information, pre-process the picked up data (e.g. by) and *reach back* to a remote fusion center using a wireless channel to transmit the required information. We refer to this setup as *reachback communication*. Based on appropriate models for the sources and the channels, the main research challenges consist of finding a suitable system architecture, optimizing distributed algorithms and communication protocols subject to the technological constraints of the sensor nodes, and providing means to search, learn and draw inferences from the wealth of gathered data.

The remainder of the chapter is organized as follows. Sect. 2.2 provides an overview of the state of the art in terms of sensor network technology. Sect. 2.3 explains the main communications challenges, and Sect. 2.4 discusses data processing issues. The chapter concludes with a brief summary and some final remarks.

2.2 Current Sensor Network Technology

Although sensor systems and technologies have been the focus of intense research efforts for several decades, wireless sensor networks have only recently begun to catch the attention of the electrical engineering and computer science communities. It is fair to say that developers of hardware platforms for distributed sensing have been consistently a few steps ahead of software programmers and protocol engineers, in that there already exist several different sensor networking prototypes available to the user, but only a rather limited number of widespread, real-world applications.

In terms of hardware development, the state of the art is well represented by a class of multi-purpose sensor nodes called *motes* [5], which were originally developed at UC Berkeley and are being deployed and tested by several research groups and start-up companies. Typical motes, such as MICA, MICAZ or TELOS-B, consist of a combination of different modules, namely a data acquisition card, a mote processor and a wireless interface. Examples of sensor platforms that are already available in the market include ambient light, barometric pressure, GPS, magnetic field, sound, photo-sensitive light, photo resistor, humidity and temperature. The mote processor performs all of the information processing tasks, from analog-to-digital conversion up to coding and upper layer protocol functions. For computation purposes, motes rely on a dedicated processor and 512 kb of non-volatile memory. The wireless module operates at radio frequencies within the available ISM bands, which can vary from slightly over 300 MHz to almost 2500 MHz. Motes can come in different sizes, depending mostly on the chosen type of energy supply, since all other components have already reached an impressive degree of miniaturization.

Another example of prototype sensor nodes was developed by the Free University of Berlin. The so called Embedded Sensor Board (ESB) includes the same kind

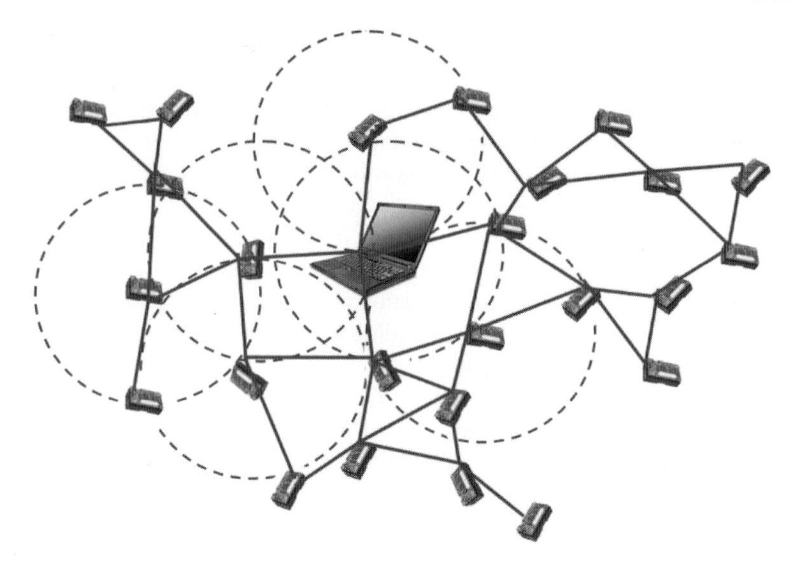

Fig. 2.2 Example of a wireless sensor network based on the Berkeley mote platform. The circles represent the transmission range of each node, which are connected by radio links (represented by edges) to all other nodes within range

of sensors, but offers very low energy consumption. In idle mode and in active mode an ESB requires 8 μA and about 10 mA, respectively, which, assuming ordinary AAA batteries and average transmission rates of 0.8 bytes per second, results in a network lifetime of 5 to 17 years.

In most of the currently available implementations, the sensor nodes are controlled by module-based operating systems such as TinyOS [30] and programming languages like nesC [9] or TinyScript/Maté [17].

It is fair to say that the programming models underlying most of these tools present one or more of the following drawbacks:

1. The model does not provide a rigorous model (or a calculus) of the sensor network at the programming level, which would allow for a formal verification of the correctness of programs, among other useful analyses.
2. The model does not provide a global vision of a sensor network application, as a specific distributed application, making it less intuitive and error prone for programmers.
3. The model requires the programs to be installed on each sensor individually, something unrealistic for large sensor networks.
4. The model does not allow for dynamic reprogramming of the network.

Recent practical middleware constructions such as Deluge [13] and Agilla [8] address a few of these drawbacks by providing higher level programming abstractions on top of TinyOS, including massive code deployment. Nevertheless, we are still far from a comprehensive programming solution with strong formal support and analytical capabilities. Beyond meeting the challenges of network-wide programming and code deployment, the ideal model should be capable of producing quantitative information on the amount of resources required by sensor network programs, and protocols and also of providing the necessary tools to prove their correctness. A proposal along these lines is presented in [28].

2.3 Communication Aspects

2.3.1 General Models

To understand the fundamental communication principles behind distributed sensing, and thus design efficient and reliable sensor networks, it is vital to find suitable models for the observed physical processes, the topology of the network, and the communications channels among other aspects. Many of the existing models share one or more of the following features:

1. *Correlated Observations*: If we have a large number of nodes sensing a physical process within a confined area, it is reasonable to assume that their measurements are correlated. This correlation may be exploited for efficient encoding/decoding.
2. *Cooperation among Nodes*: To transmit data to the remote receiver, the sensor nodes may organize themselves in a suitable network topology to exchange information over the wireless medium and increase their efficiency or flexibility through cooperation.
3. *Channel Interference*: If multiple sensor nodes use the wireless medium at the same time (either for control messages or data packets), their signals will necessarily interfere with each other. Consequently, reliable communication in a sensor network requires a set of rules that control (or exploit) the interference in the wireless medium.
4. *Energy Consumption*: Since the operation of the sensor nodes is severely limited by the available battery power, the energy consumption is a very important parameter governing the lifetime and efficiency of wireless sensor networks.

Example 2. As a typical instance, consider the system model for a single-hop sensor network illustrated in Fig. 2.3. Each sensor k observes at time t continuous real-valued data samples $u_k(t)$, with $k = 1, 2, \ldots, M$. For simplicity, we assume that the M sensor nodes are placed randomly on the unit square and consider only the spatial correlation of measurements and not their temporal dependence. Thus, we drop the time variable t and consider only one time step. The sample vector $\mathbf{u} = (u_1 \, u_2 \cdots u_M)^T$ at any given time t is assumed to be one realization of an M-dimensional Gaussian random variable, whose PDF $p(\mathbf{u})$ is given by $\mathcal{N}(\mathbf{0}_M, \mathbf{R})$ with

$$\mathbf{R} = \begin{bmatrix} 1 & \rho_{1,2} & \cdots & \rho_{1,M} \\ \rho_{2,1} & 1 & \cdots & \rho_{2,M} \\ \vdots & \vdots & \ddots & \vdots \\ \rho_{M,1} & \rho_{M,2} & \cdots & 1 \end{bmatrix}.$$

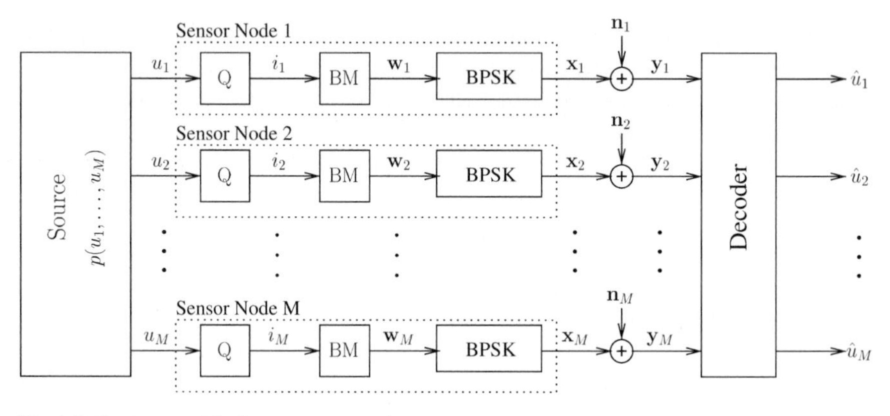

Fig. 2.3 System model of a sensor network

Gaussian models for capturing the spatial correlation between sensors at different locations are discussed in [25], whereas reasonable models for the correlation coefficients $\rho_{i,j}$ of physical processes unfolding in a field can be found in [6].

Furthermore, we assume that the sensors are randomly placed in a unit square according to a uniform distribution and that the correlation $\rho_{i,j}$ between sensor i and j decays exponentially with their Euclidean distance $d_{i,j}$, i.e., $\rho_{i,j} = \exp(-\beta \cdot d_{i,j})$, where β is a positive constant. Note that this correlation structure, which we deem to be a reasonable abstraction of the physical measurements picked up locally by a number of scattered sensors, is only one of many available source models.

The model assumes that the sensors are "cheap" devices consisting of a scalar quantizer, a bit mapper and a modulator. Each sensor k quantizes u_k to the index $i_k \in \mathcal{L} = \{1, 2, \ldots, 2^Q\}$, representing Q bits. The modulator maps i_k to a tuple $\mathbf{x}_k$ of channel symbols, which are transmitted to the remote receiver using a simple BPSK modulation.

Since in many applications sensors must transmit some data to the central receiver simultaneously, reservation-based medium access protocols such as TDMA or FDMA are a reasonable choice for this type of single-hop networks. Thus, we assume that the reachback channel is virtually interference-free, i.e., the joint PDF $p(\mathbf{y}_1, \ldots, \mathbf{y}_M | \mathbf{x}_1, \ldots, \mathbf{x}_M)$ factors into $\prod_{k=1}^{M} p(\mathbf{y}_k | \mathbf{x}_k)$. Frequently, the reachback channel appears as an array of additive white Gaussian noise channels with noise variance σ^2, i.e., the channel outputs are given by $\mathbf{y}_k = \mathbf{x}_k + \mathbf{n}_k$ after demodulation.

In most cases, the sensor nodes are not connected directly to the base station and therefore multi-hop communication becomes imperative. The network topology is usually dynamic and depends largely on (a) the chosen form of deployment, e.g. in a pre-determined grid or randomly placed; (b) the characteristics of the physical environment and the communication channels; (c) energy constraints; and (d) the reliability and robustness of sensor nodes. A common mathematical model for sensor network connectivity is the random geometric graph. A random geometric graph, $G(n, r)$, is formed as follows: place n nodes uniformly at random onto the surface of a two-dimensional unit square or torus, and connect nodes within Euclidean distance r of each other. Because an edge (v_i, v_j) exists if and only if the distance between nodes v_i and v_j is less than r, the edge (v_i, v_j) represents the connectivity between sensor i and sensor j.

Arguably the most important factor determining the lifetime of a wireless sensor network, i.e. the maximum period of time under which the functioning nodes can guarantee the connectivity of the network, is power preservation. In other words, unless the algorithms and communication protocols used by the sensors to gather and convey information consume a minimal amount of power, one by one the sensors will exhaust their resources and the network will cease to guarantee enough paths over which each sensor can send its collected data to the intended destination. One way to save power is to exploit the correlations in the sensor data and transmit less bits using distributed data compression algorithms, as will be explained in Sect. 2.4. This form of distributed signal processing is also beneficial with respect to the routing of messages within the network, as shown in [25]. Another option is to limit the operation of each sensor node to a very small duty cycle, which can decisively reduce the overall power consumption but renders the network more dynamic, and thus more difficult to analyze and manage.

It is fair to say that the behavior of dynamic sensor networks is not yet well understood and further progress is likely to require recent tools of complex network analysis based on more elaborate random graph constructions. Such models could be used to determine the fundamental performance limits of wireless sensor networks and constructive strategies to achieve them.

2.3.2 Routing and Data Gathering Protocols

Once the sensor nodes collect measurements from the physical environment it is necessary to transmit the relevant data to one or more data collection points, possibly over multiple hops across the network. This process, which is commonly referred to as *data gathering*, is intimately related to the message routing problem in communication networks. A key feature which distinguishes sensor networks in this context is that the data to be conveyed among the nodes do not need to be an exact representation of the values they measured. In fact, depending on the amount of allowable in-network processing, the messages that are routed through the network are likely to contain compressed versions, subsets, combinations or functions of the measurements—just enough data for the fusion center to be able to obtain the information that is relevant for the particular application.

If the data collection point is within the range of every sensor, as in the example of Fig. 2.3, the data-gathering problem may be solved by direct communication between the base station and all the nodes. In general, this solution is only possible in relatively small environments, otherwise the transmit power and consequently the energy consumption would be too high and limit severely the sensor network's lifetime. Similar to all other networks, flooding the network is effective at disseminating information but the resulting broadcast storms generate interference and waste energy, thus compromising the reliability of the network. An alternative offered by several existing data-gathering protocols is for the sensor nodes to relay the sent messages over multiple hops until they reach their destiny. This is the case, for instance, with the Low-Energy Adaptive Clustering (LEACH) protocol [12] which reduces the number of nodes that communicate directly with the base station by forming a small number of clusters in a self-organized and dynamic way. In each cluster, a randomly chosen head is responsible for fusing the data locally from its neighboring nodes and transmitting it to the data collection point. Cluster heads make about 5% of the sensor nodes, which accounts for the protocol's efficiency. A similar approach is proposed by the Power-Efficient Gathering in Sensor Information Systems (PEGASIS) protocol [18], where the sensor nodes form a chain, such that each node communicates only with the closest possible neighbor. The chain can result from a greedy distributed algorithm or from a central solution computed by the base station and sent via broadcast to all the nodes.

The Power Efficient Data Gathering and Aggregation Protocol (PEDAP, [21]) improves on the previous data-gathering protocols by assuming that the location of each sensor node is known at the base station, thus enabling the computation of optimal routes using a minimum spanning tree algorithm. If a node fails, the base station recomputes the routing tables and sends this information to all nodes in the network.

A somewhat different approach is offered by Directed Diffusion [14], in which the data collection point broadcasts queries with a rich set of attribute-value pairs describing the desired data. These queries are cached by intermediate nodes and once the data are available it is routed using a gradient towards the data collection point.

The aforementioned data-gathering protocols are only a small subset of currently available proposals. As with many topics in sensor networks, the choice of communication protocol is highly dependent on the characteristics and demands of the envisioned application.

2.3.3 Security Considerations

The processing constraints of sensor nodes make it hard to implement standard cryptographic primitives used in today's communication networks. Sensor network security thus remains a challenging problem, for which there are not yet many technical solutions.

In general terms, typical attacks on wireless networks can be divided into passive and active attacks. Eavesdropping and traffic analysis can be considered passive attacks, because the unauthorized party gains access to a network asset but does not modify its content. Active attacks are those in which the unauthorized party modifies messages or data streams, for instance:

- *Message Modification*: the attacker deletes, adds, changes or re-orders messages.
- *Denial-of-Service*: the attacker inhibits legitimate communications.
- *Masquerading*: the attacker pretends to be an authorized user to gain access to resources.
- *Replay*: the attacker retransmits previously stored messages at a later point in time.

Thus, a central task in any security scheme for sensor networks is to detect malicious behavior within the network. This can be achieved by having each node monitor the transmissions of its neighbors and reporting any observation that arouses suspicion. Typical attacks specific to sensor networks include the generation of false traffic messages to compromise routing and data gathering efforts, dropping of messages that ought to be relayed, presenting multiple identities to other nodes in the network, pretending to be a cluster head, jamming other nodes, impersonating the base station, creating *wormholes* to tunnel the data to an unforeseen destination, among many others.

As an example, SPINS (Security Protocols for Sensor Networks) is a protocol suite designed to meet specific security criteria [22], including data confidentiality, two-party data authentication, evidence of data freshness and authenticated broadcast.

2.4 Distributed Compression and In-Network Computation

2.4.1 Distributed Compression

In distributed sensing scenarios, where data from a large number of low-complexity and power-restricted sensors have to be gathered, efficient yet simple data compression techniques are one of the key-factors in enabling high spatial density and long lifetime. Assuming that both factors are affected by the number of communication operations required—which is surely true for scenarios where a common channel has to be shared and/or the power consumption of the sensor is highly affected by the time of transmitter utilization—any reduction in the amount of data that needs to be transmitted over the wireless medium will have a decisive impact on the overall power consumption of the sensor nodes and the longevity of the sensor network. Since the measurements picked up by the sensor nodes will in general be correlated, it is only natural to consider data compression schemes that remove the inherent redundancy with manageable complexity. One way is to exploit the statistics of the observed physical process and allow the sensors to encode their measurements independently of each other, i.e. without any cooperation/data exchange. The basic principle behind this idea is illustrated in the following two examples.

Example 3 (From [33]). Assume that Alice must communicate an even number x to Bob and that Bob receives as side information one of the neighbors of x. If Alice has access to Bob's side information, then she only needs to send one bit, e.g., indicating whether x is above or below the known neighbor. What if Alice does not know the side information available to Bob? A possible solution for this problem is to place all possible even numbers into two bins: one bin for all multiples of four and one bin for all the others. By sending the bin index (0 or 1) to Bob, Alice is able to achieve the exact same efficiency (1 bit) even without access to the side information.

Example 4. Let U_1 and U_2 be two correlated sources of information, which output binary triplets differing at most in one bit, e.g., 000 and 010 or 101 and 001. Assuming that U_1 must be communicated by the transmitter and that U_2 is available as side information at the decoder, how should we encode the former, such that perfect reconstruction is possible with a minimum amount of transmitted bits? Notice that there are only two bits of uncertainty, i.e. enough to indicate in which bit U_1 and U_2 differ, provided of course that U_2 is known at the encoder. Interestingly enough, the same coding efficiency can still be achieved even if U_2 is not known at the encoder. Here is the key idea: it is not necessary for the code to differentiate between U_1 triplets that differ in three positions, because the decoder will count the number of bits that are different from U_2 and only one of the two possible U_1 triplets will be one bit away. Thus, by putting the eight possible realizations of U_1 in four bins and guaranteeing that the elements in one bin differ in three bit positions, we can rest assured that U_1 will be perfectly reconstructed at the decoder.

From these examples we conclude that it is possible to eliminate the redundancy in sensor data even when each sensor node does not know the observations of its peers. In general terms this so-called *distributed source coding* problem can be formulated as follows. Assume that two information sources, U_1 and U_2, are to be processed by a joint encoder and transmitted to a common destination over two noiseless channels, as shown in Fig. 2.4. In general, $p(u_1u_2) \neq p(u_1)p(u_2)$, such that the messages produced by U_1 and U_2 at any given point in time are statistically dependent—we refer to U_1 and U_2 as *correlated sources*. Since the channels do not introduce any errors, we may ask the following question: at what rates R_1 and R_2 can we transmit information generated by U_1 and U_2 with arbitrarily small probability of error? Not surprisingly, since we have a common encoder and a common decoder, this problem reduces to the classical point-to-point problem and the solu-

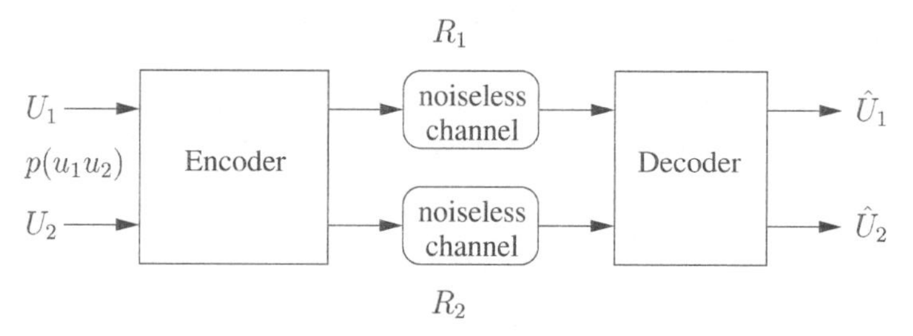

Fig. 2.4 Joint encoding of correlated sources

tion follows naturally from Shannon's source coding theorem: the messages can be perfectly reconstructed at the receiver if and only if

$$R_1 + R_2 > H(U_1U_2),$$

i.e. the sum rate must be greater than the joint Shannon entropy of U_1 and U_2.

The problem becomes considerably more challenging if, instead of a joint encoder, we have two *separate* encoders, as shown in Fig. 2.5. Here, each encoder observes only the realizations of the one source it is assigned to and does not know the output symbols of the other source. In this case, it is not immediately clear which encoding rates guarantee perfect reconstruction at the receiver. If we encode U_1 at rate $R_1 > H(U_1)$ and U_2 at rate $R_2 > H(U_2)$, then the source coding theorem guarantees once again that arbitrarily small probability of error is possible. But, in this case, the sum rate amounts to $R_1 + R_2 > H(U_1) + H(U_2)$, which in general is greater than the joint entropy $H(U_1U_2)$.

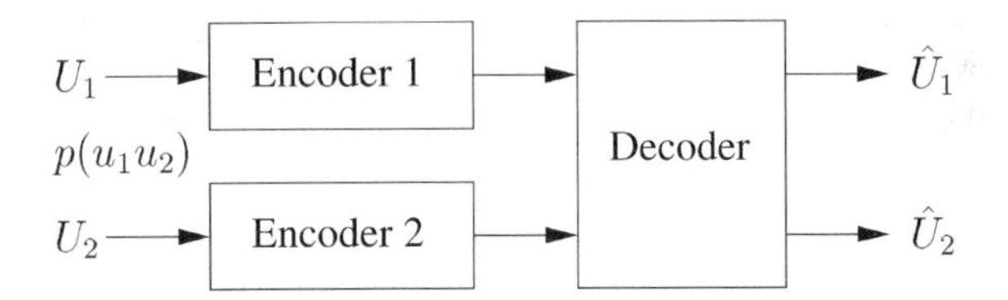

Fig. 2.5 Separate encoding of correlated sources (*Slepian–Wolf problem*). The noiseless channels between the encoders and the decoder are omitted

In their landmark paper [29], Slepian and Wolf come to a surprising conclusion: the sum rate required by two separate encoders is the same as that required by a joint encoder, i.e. $R_1 + R_2 > H(U_1U_2)$ is sufficient for perfect reconstruction to be possible. In other words, there is no loss in overall compression efficiency due to the fact that the encoders can only observe the realizations of the one source they have been assigned to. However, it is important to point out that the decoder does require a minimum amount of rate from each encoder, specifically the average remaining uncertainty about the messages of one source given the messages of the other source, i.e. $H(U_1|U_2)$ and $H(U_2|U_1)$. The set of achievable compression rates is thus fully characterized by the following theorem.

Theorem 1 (Slepian–Wolf Theorem, [29]). *Let (U_1U_2) be two correlated sources drawn i.i.d. $\sim p(u_1u_2)$. The compression rates (R_1, R_2) are achievable if and only if*

$$R_1 \geq H(U_1|U_2),$$
$$R_2 \geq H(U_2|U_1),$$
$$R_1 + R_2 \geq H(U_1U_2).$$

Inspired by the seminal work of Slepian and Wolf, several authors have contributed with distributed source coding solutions (see e.g. [4] and references therein).

Focusing on scalar quantization, Flynn and Gray [7] provided one of the first practical approaches to constructing distributed source codes for two continuous-valued sources. The basic idea behind this approach is to reuse the indices of a high-resolution quantizer, such that the overall end-to-end distortion after joint decoding is minimized. Pradhan and Ramchandran presented in [23] a method called distributed source coding using syndromes (DISCUS), based on channel codes with good distance properties, where the set of possible codewords is partitioned into co-sets and only the co-set's syndrome and not the actual codeword is transmitted to the decoder. This method, originally considered for an asymmetric scenario where the decoder has access to side information, was recently extended to the asymmetric case [24]. An alternative approach for the asymmetric scenario with side information at the decoder was provided by Servetto in [27], where a constructive approach for Gaussian sources based on linear codes and nested lattices was presented. Cardinal and Van Asche [31] as well as Rebello-Monedero and Girod [10] focused on the optimization of the quantization stage and proposed design algorithms for multi-terminal quantizers. Beyond these contributions, highly evolved iterative channel coding techniques such as low density parity check (LDPC) and turbo codes have been applied to the distributed source coding problem [4], reaching the fundamental limits of Slepian and Wolf [29] for lossless distributed source coding, as well as the rate-distortion function with side information of Wyner and Ziv [32].

In [1], a network architecture that consists of separate distributed compression, channel coding and routing modules is shown to be information-theoretically optimal for general sensor networks, in which interference is dealt with by means of orthogonal accessing (e.g. TDMA or FDMA). Alternatively, the sensor nodes can encode their data in a collaborative way, which might be more practicable when data is communicated within a multi-hop framework, in which case additional information from forwarded packets may be retrieved and used (as side information) for data compression [25].

If the correlation between all the sensor nodes is to be taken into consideration, the complexity of the information processing algorithms can quickly become a major bottleneck in system performance. For example, to fully exploit the correlation in the data across the entire network, each node would have to know the joint statistics of all nodes and use complex distributed codes, two requirements which may turn out to be impractical for large-scale sensor networks. Moreover, it can be shown that for a large-scale sensor network with hundreds of nodes, the optimal decoder based on minimum mean square error (MMSE) estimation is unfeasible, because its complexity grows exponentially with the number of sensors in the network [2].

One way to tackle this problem is to shift the system complexity from the sensor nodes to the receiver by designing a sensor network with very simple encoders (e.g. a scalar quantizer and a modulator) and a decoder of increased yet manageable complexity. In [2] this is achieved by means of a factor graph [16] that models the correlation between the sensor signals in a flexible way depending on the targeted decoding complexity and the desired reconstruction fidelity. Then, based on this factor graph, the sum-product algorithm [16] is used to estimate the transmitted data. By choosing factor graph in an appropriate way it is possible to make the overall decoding complexity grow linearly with the number of nodes. Naturally, the performance of the decoding algorithm depends heavily on the accuracy of the chosen factor graph as a model for the correlation in the sensor data. Distributed source codes and clustering algorithms designed along these lines are presented in [19].

2.4.2 Distributed Computation and Inference

Depending on the sensor networks' application, distributed inference and in-network computation principles can be used to efficiently perform distributed data analysis tasks. In scenarios where one is interested in some intermediate measure (e.g. the median or the maximum value) obtained by one or a collection of sensors, instead of the measured data set itself, only the intermediate measure has to be communicated to the destination. For example, consider the possible application of fire detection, where one does not need the complete data set of all temperature measurements from all sensors, but only spatiotemporal temperature gradients or even only the maximum of those values. Exploiting distributed inference and in-network computation principles, the amount of data to be communicated along the network to the access point can thus be reduced considerably and the networks' lifetime as well as its scalability can be improved.

The main channel here is that of ensuring that the communication channels carry enough information from multiple sensor nodes to enable the reliable computation of some functional of interest, in many instances given the data already available at the destination. At the current stage, we are still far from providing general answers to this problem, in fact this field is largely unexplored territory.

As a guiding principle, it would be useful to know the theoretical limits for the end-to-end distortion of the functional's estimate in relation to the information available to the estimator. The performance of different estimation algorithms could then be evaluated against the ultimate rate-distortion bounds. Unfortunately, obtaining such bounds appears to be beyond the ability of current information-theoretic techniques. In the following, we discuss some of the open issues.

A first approach towards the problem of distributed computation over noisy channels would be to encode the samples separately using multi-terminal source codes [29,3] with some distortion and transmit them reliably over the noisy channels using point-to-point channel codes. The decoder could start by recovering the sent data and computing the desired functional. To be able to compute the fundamental performance limits of this approach, it would be necessary to know the rate-distortion region for separate compression of multiple correlated sources, a problem that has been open for 30 years (for recent progress, see [26]).

A different approach, which is likely to outperform the previous one in some instances, would be to use joint or separate source and channel codes that ensure that the decoder has sufficient statistics from which to compute the target functional. For noiseless channels, Körner and Marton characterized the rates required to reconstruct the modulo-two sum of two correlated uniform binary sources [15], and show that it is possible to save rate by using adequate source codes. More recently, Orlitsky and Roche determined how many bits must be transmitted by an encoder observing X_1 for a decoder with side information X_2 to be able to compute $f(X_1, X_2)$ [20]. The main difference is that here the side information is uncoded and available at the decoder. Generalizing the results to a more realistic sensor network model is most definitely a very challenging task.

2.5 Summary and Concluding Remarks

In this chapter, we provided a brief introduction to wireless sensor networks as an enabling technology for distributed sensing, computing and data gathering. Perhaps the most striking observation to make is that the design of system architecture, algorithms and protocols is imminently application-driven and therefore general remarks on the nature of sensor networks and their fundamental mechanisms are necessarily limited in scope. The same observation applies to basic models, data gathering strategies, learning and inference algorithms and security considerations. If there is one common characteristic among many different classes of sensor networks, it is the central concern with low energy consumption and the consequent need for power-aware algorithms and protocols. In many cases, data-centric design and full use of the potential for cooperation between sensor nodes is key towards achieving this goal.

Acknowledgements The author gratefully acknowledges discussions with his collaborators Sergio D. Servetto, Gerhard Maierbacher, Luísa Lima, Paulo Oliveira, Pedro Brandão, Michael Tuechler, Luís Lopes, Francisco Martins, and Miguel Santos Silva.

References

[1] J. Barros, S.D. Servetto, Network information flow with correlated sources. IEEE Transactions on Information Theory, 52(1):155–170, 2006.
[2] J. Barros, M. Tuechler, Scalable decoding on factor trees: a practical solution for sensor networks. IEEE Transactions on Communications, 54:284–294, 2006.
[3] T. Berger, Multiterminal source coding. In: G. Longo (Ed.), The Information Theory Approach to Communications. Springer, Berlin, 1978
[4] S. Cheng, Z. Xiong, A.D. Liveris, Distributed source coding for sensor networks. IEEE Signal Processing Magazine, September 2004.
[5] D.E. Culler, H. Mulder, Smart Sensors to Network the World. Scientific American, 2004.
[6] C.R. Dietrich, G.N. Newsam, Fast and exact simulation of stationary Gaussian processes through circulant embedding of the covariance matrix. SIAM Journal on Scientific Computing, 18(4):1088–1107, 1997.
[7] T.J. Flynn, R.M. Gray, Encoding of correlated observations. IEEE Transactions on Information Theory, IT-33(6):773–787, 1987.

[8] C.-L. Fok, G.-C. Roman, C. Lu, Rapid development and flexible deployment of adaptive wireless sensor network applications. In: Proceedings of the 24th International Conference on Distributed Computing Systems (ICDCS'05), pp. 653–662. IEEE, June 2005.
[9] D. Gay, P. Levis, R. von Behren, M. Welsh, E. Brewer, D. Culler, The nesC language: a holistic approach to network embedded systems. In: ACM SIGPLAN Conference on Programming Language Design and Implementation (PLDI), 2003.
[10] B. Girod, D. Rebollo-Monedero, R. Zhang, Design of optimal quantizers for distributed source coding. In: Proceedings of the Data Compression Conference (DCC03), 2003.
[11] R. Grisso, M. Alley, P. McClellan, D. Brann, S. Donohue, Precision farming: a comprehensive approach. Technical Report, Virginia Tech, 2002. Available from http://www.ext.vt.edu/pubs/bse/442-500/442-500.html
[12] W.R. Heinzelman, J. Kulik, H. Balakrishnan, Adaptive protocols for information dissemination in wireless sensor networks. In: MobiCom '99: Proceedings of the 5th Annual ACM/IEEE International Conference on Mobile Computing and Networking, pp. 174–185, New York, NY, USA. ACM Press, 1999.
[13] J.W. Hui, D. Culler, The dynamic behavior of a data dissemination protocol for network programming at scale. In: Proceedings of the 2nd International Conference on Embedded Networked Sensor Systems, pp. 81–94. ACM Press, 2004.
[14] C. Intanagonwiwat, R. Govindan, D. Estrin, J. Heidemann, F. Silva, Directed diffusion for wireless sensor networking. IEEE/ACM Transactions on Networks, 11(1):2–16, 2003.
[15] J. Körner, K. Marton, How to encode the modulo 2 sum of two binary sources. IEEE Transactions on Information Theory, IT-25:219–221, 1979.
[16] F.R. Kschischang, B. Frey, H.-A. Loeliger, Factor graphs and the sum-product algorithm. IEEE Transactions on Information Theory, 47(2):498–519, 2001.
[17] P. Levis, D. Culler, Maté: a tiny virtual machine for sensor networks. SIGOPS Operating Systems Review, 36(5):85–95, 2002.
[18] S. Lindsey, C.S. Raghavendra, Pegasis: power-efficient gathering in sensor information systems. In: IEEE Aerospace Conference, vol. 3, pp. 1125–1130, 2002.
[19] G. Maierbacher, J. Barros, Source-optimized clustering for distributed source coding. In: Proceedings of the IEEE Global Telecommunications Conference (GLOBECOM'06), San Francisco, USA, 2006.
[20] A. Orlitsky, J.R. Roche, Coding for computing. IEEE Transactions on Information Theory, 47(3):903–917, 2001.
[21] H. Özgür Tan, I. Körpeoglu, Power efficient data gathering and aggregation in wireless sensor networks. SIGMOD Record, 32(4):66–71, 2003.
[22] A. Perrig, R. Szewczyk, J.D. Tygar, V. Wen, D.E. Culler, SPINS: security protocols for sensor networks. Wireless Networks, 8(5):521–534, 2002.
[23] S.S. Pradhan, K. Ramchandran, Distributed source coding using syndromes (DISCUS): design and construction. In: Proc. IEEE Data Compression Conf. (DCC), Snowbird, UT, 1999.
[24] S. Sandeep Pradhan, K. Ramchandran, Generalized coset codes for distributed binning. IEEE Transactions on Information Theory, 51:3457–3474, 2005.
[25] A. Scaglione, S.D. Servetto, On the interdependence of routing and data compression in multi-hop sensor networks. In: Proc. ACM MobiCom, Atlanta, GA, 2002.
[26] S.D. Servetto, Multiterminal source coding with two encoders–i: a computable outer bound, 2006. IEEE Transactions on Information Theory; Submitted, April 2006, Revised, November 2006.
[27] S.D. Servetto, Lattice quantization with side information: codes, asymptotics, and applications in sensor networks. IEEE Transactions on Information Theory, 53(2):714–731, 2007.
[28] M.S. Silva, F. Martins, L. Lopes, J. Barros, A calculus for sensor networks. Available from http://arxiv.org/abs/cs.DC/0612093, December 2006.
[29] D. Slepian, J.K. Wolf, Noiseless coding of correlated information sources. IEEE Transactions on Information Theory, IT-19(4):471–480, 1973.
[30] The TinyOS Documentation Project. Technical Report. Available at http://www.tinyos.net.

[31] G. Van Assche Jean Cardinal, Joint entropy-constrained multiterminal quantization. In: Proceedings of the International Symposium on Information Theory, Lausanne, Switzerland, 2002.
[32] A.D. Wyner, J. Ziv, The rate-distortion function for source coding with side information at the decoder. IEEE Transactions on Information Theory, IT-22(1):1–10, 1976.
[33] R. Zamir, S. Shamai, U. Erez, Nested codes: an algebraic binning scheme for noisy multiterminal networks. IEEE Transactions on Information Theory, 2002. To appear. Available from http://www.eng.tau.ac.il/ zamir/.

Chapter 3
Data Stream Processing

João Gama and Pedro Pereira Rodriques

3.1 Introduction

The rapid growth in information science and technology in general and the complexity and volume of data in particular have introduced new challenges for the research community. Many sources produce data continuously. Examples include sensor networks, wireless networks, radio frequency identification (RFID), customer click streams, telephone records, multimedia data, scientific data, sets of retail chain transactions etc. These sources are called data streams. A data stream is an ordered sequence of instances that can be read only once or a small number of times using limited computing and storage capabilities. These sources of data are characterized by being open-ended, flowing at high-speed, and generated by non stationary distributions in dynamic environments.

What distinguishes current data from earlier one is automatic data feeds. We do not just have people who are entering information into a computer. Instead, we have computers entering data into each other [25]. Nowadays there are applications in which the data are modeled best as transient data streams instead of as persistent tables. Examples of applications include network monitoring, user modeling in web applications, sensor networks in electrical networks, telecommunications data management, prediction in stock markets, monitoring radio frequency identification etc. In these applications it is not feasible to load the arriving data into a traditional data base management system (DBMS) and traditional DBMS are not designed to directly support the continuous queries required by these applications [3]. Carney

J. Gama
LIAAD–INESC Porto L.A. and Faculty of Economics, University of Porto, Rua de Ceuta, 118-6, 4050-190 Porto, Portugal
e-mail: jgama@fep.up.pt

P.P. Rodriques
LIAAD–INESC Porto L.A. and Faculty of Sciences, University of Porto, Rua de Ceuta, 118-6, 4050-190 Porto, Portugal
e-mail: pprodriques@fc.up.pt

et al. [6] pointed out the significant differences between data bases that are passive repositories of data and data bases that actually monitor applications and alert humans when abnormal activity is detected. In the former, only the current state of the data is relevant for analysis. Humans initiate queries, usually one-time, predefined queries. In the latter, data come from external sources (e.g., sensors), and require processing historic data. For example, in monitoring activity, queries should run continuously. The answer to a continuous query is produced over time, reflecting the data seen so far. Moreover, if the process is not strictly stationary (as most of real-world applications), the target concept could gradually change over time. For example, the type of abnormal activity (e.g., attacks in TCP/IP networks, frauds in credit card transactions etc.) changes over time.

Organizations use decision support systems to identify potential useful patterns in data. Data analysis is complex, interactive, and exploratory over very large volumes of historic data, eventually stored in distributed environments. Traditional pattern discovery process requires online ad-hoc queries, not previously defined, that are successively refined. Nowadays, given the current trends in decision support and data analysis, the computer plays a much more active role, by searching hypotheses, evaluating and suggesting patterns. Due to the exploratory nature of these queries, an exact answer may not be required. A user may prefer a fast approximate answer. Range queries and selectivity estimation (the proportion of tuples that satisfy a query) are two illustrative examples where fast but approximate answers are more useful than slow and exact ones.

Sensor networks are distributed environments producing multiple streams of data. We can consider the network as a distributed database we are interested in querying and mining. In this chapter we review the main techniques used for query and mining data streams that are of potential use in sensor networks.

In Sect. 3.2 we refer to the data stream models and identify its main research challenges. Section 3.3 presents basic stream models. Section 3.4 present basic stream algorithms for maintaining synopsis over data streams. Section 3.5 concludes the chapter and points out future directions for research.

3.2 Data Stream Models

What makes data streams different from the conventional relational model? A key idea is that operating in the data stream model does not preclude the use of data in conventional stored relations. Some relevant differences include [3]:

- Data elements in the stream arrive online.
- The system has no control over the order in which data elements arrive, either within a data stream or across data streams.
- Data streams are potentially unbounded in size.
- Once an element from a data stream has been processed it is discarded or archived. It cannot be retrieved easily unless it is explicitly stored in memory, which is small relative to the size of the data streams.

In the stream model [25] the input elements $a_1, a_2, \ldots, a_j, \ldots$ arrive sequentially, item by item and describe an underlying function A. Stream models differ on how a_i describes A. We can distinguish between:

- Insert Only Model: once an element a_i is seen, it cannot be changed.
- Insert-Delete Model: elements a_i can be deleted or updated.
- Accumulative Model: each a_i is an increment to $A[j] = A[j-1] + a_i$.

Examples of stream applications for the three models include time-series generated by sensor networks (Insert Only Model), radio frequency identification (Insert-Delete Model), monitoring the Max (or Min) of the sum of quantities transactioned per entity (Accumulative Model).

3.2.1 An Illustrative Problem

A simple problem that clearly illustrates the issues in stream process is finding the maximum value (MAX) or the minimum value (MIN) in a sliding window over a sequence of numbers. When we can store in memory all the elements of the sliding window, the problem is trivial and we can find the exact solution. When the size of the sliding window is greater than the available memory, there is no algorithm that provides an exact solution, because it is not possible to store all the data elements. For example, suppose that the sequence is monotonically decreasing and the aggregation function is MAX. Whatever the window size, the first element in the window is always the maximum. As the sliding window moves, the exact answer requires maintaining all the elements in memory.

3.2.2 Research Issues in Data Streams Management Systems

Babcock et al. [3] defined *blocking query operators* as query operators that are unable to produce the first tuple of the output before they have seen the entire input. Aggregating operators, like SORT, SUM, COUNT, MAX, MIN, are blocking query operators. In the stream setting, continuous queries using block operators are problematic. The semantics of these operators in the stream problem setting is an active research area. For example, Raman et al. [26] proposed an approximate online version for *Sort*. Fang et al. [12] studied approximate algorithms for *iceberg queries* using data synopsis. Iceberg queries compute aggregate functions (such as COUNT, SUM) over an attribute to find aggregate values above some specified threshold. They process a relative large amount of data to produce results where the number of above-threshold results is often very small (the tip of an iceberg).

From the point of view of a data stream management system, several research issues emerge [3]. These issues have implications in data stream management systems, like:

- Approximate query processing techniques to evaluate queries that require an unbounded amount of memory.
- Sliding window query processing, both as an approximation technique and as an option in the query language.
- Sampling to handle situations where the flow rate of the input stream is faster than the query processor.
- The meaning and implementation of blocking operators (e.g., aggregation and sorting) in the presence of unending streams.

These types of queries require techniques for storing summaries or synopsis information about previously seen data. There is a trade-off between the size of summaries and the ability to provide precise answers.

The time dimension is an important characteristic in streams. For example, in time window models, it indicates when a specific measure is valid. In single streams, the meaning of a time-stamp attribute, indicating when the tuple has arrived, has clear semantics. When a query involves multiple streams, the semantics of time stamps is much less clear. For example, what is the time stamp of an output tuple? Research in temporal databases propose several methods to deal with time, including temporal types [27,5].

A large number of applications are distributed in nature. Sensors in electrical networks are geographically distributed. Data collected from these sensors have not only a time dimension but also a space dimension. Furthermore, sensors are resource limited. While the time dimension is critical for synchronization aspects, the space dimension makes bandwidth and battery power critical aspects as well.

A summary of the differences between traditional and stream data processing is presented in Table 3.1.

Table 3.1 Differences between traditional and stream data processing

	Traditional	Stream
Number of passes	Multiple	Single
Processing Time	Unlimited	Restricted
Memory Usage	Unlimited	Restricted
Type of Result	Accurate	Approximate
Distributed	No	Yes

3.3 Basic Streaming Methods

Data streams are unbounded in length. However, this is not the only problem. The domain of the possible values of an attribute is also very large. A typical example is the domain of all IP addresses on the Internet.[1] It is so huge that complete stor-

[1] Actually, in the new standard protocol for the Internet, IPv6, an IP address is a number defined by 128 bits [20]. Thus, the number of all IP addresses is 2^{128}, around 600,000 addresses for every square nanometer of the entire Earth's surface.

age is an intractable problem. The execution of queries that reference past data also becomes problematic. These are situations in which stream methods are applicable and become relevant. Stream methods provide approximate answers using the reduced resources.

It is impractical to store all data to execute queries that reference past data. Seeking a fast but approximate answer, however, is a valid alternative. Several techniques have been developed for storing summaries or synopsis information about previously seen data. There is a trade-off between the size of summaries and the ability to provide precise answers. Datar et al. [11] presented simple counting problems that clearly illustrate the issues in data stream research:

1. Given a stream of bits (0's and 1's), maintain a count of the number of 1's in the last N elements seen from the stream.
2. Given a stream of elements that are positive integers in the range $[0, \dots, R]$, maintain at every instance the sum of the last N elements seen from the stream.
3. Find the number of distinct values in a stream of values with domain $[1, \dots, N]$.

All these problems have an exact solution if we have enough memory to store all the elements in the sliding window. That is, the exact solution requires $O(N)$ space. Suppose we have restricted memory. How can we solve these problems using less space than $O(N)$?

Approximate answers are useful if the associated error is in an admissible boundary. We can define the approximation methods in the data streams as:

> (ϵ, δ)-approximation schema: Given any positive number $\epsilon < 1$ and $\delta < 1$ compute an estimate that, with probability $1 - \delta$, is within relative error $\leq \epsilon$.

Time and space required to compute an answer depends on ϵ and δ. Some results on tail inequalities provided by statistics are useful to accomplish this goal. The basic general bounds on the tail probability of a random variable (that is, the probability that a random variable deviates greatly from its expectation) include the Markov, and Chebyshev inequalities [24]. Two other powerful bounds derived from those are the Chernoff and Hoffding bounds.

Statistical Inequalities

An estimator is a function of the observable sample data that is used to estimate an unknown population parameter. We are particularly interested in interval estimators that compute an interval for the true value of the parameter associated with a confidence $1 - \delta$. Two types of intervals are:

- Absolute approximation: $\overline{X} - \epsilon \leq \mu \leq \overline{X} + \epsilon$, where ϵ is the absolute error; and
- Relative approximation: $(1 - \delta)\overline{X} \leq \mu \leq (1 + \delta)\overline{X}$, where δ is the relative error.

The following two results from statistical theory are useful in most of the cases.

Theorem 1 (Chernoff Bound). *Let $X_1, X_2, \dots, X_n$ be independent random variables from Bernoulli experiments. Assume that $P(X_i = 1) = p_i$. Let $X_s = \sum_{i=1}^{n} X_i$ be a random variable with expected value $\mu_s = \sum_{i=1} np_i$. Then for any $\delta > 0$*

$$P\left[X_s > (1+\delta)\mu_s\right] \le \left(\frac{e^{\delta}}{(1+\delta)^{1+\delta}}\right)^{\mu_s}. \tag{3.1}$$

From this theorem, its possible to derive the absolute error [24]:

$$\epsilon \le \sqrt{\frac{3\bar{\mu}}{n}\ln(2/\delta)}. \tag{3.2}$$

Theorem 2 (Hoeffding Bound). *Let $X_1, X_2, \ldots, X_n$ be independent random variables. Assume that each x_i is bounded, that is $P(X_i \in R = [a_i, b_i]) = 1$. Let $S = 1/n \sum_{i=1}^{n} X_i$, with the expected value $E[S]$. Then for any $\epsilon > 0$,*

$$P\left[S - E[S] > \epsilon\right] \le e^{-\frac{2n^2\epsilon^2}{R^2}}. \tag{3.3}$$

In other words, with probability $1-\delta$ the true mean is within ϵ of the sample mean. From this theorem [24], we can derive a bound on how closed the estimated mean is from the true mean after seeing n points with confidence $1-\delta$;

$$\epsilon \le \sqrt{\frac{R^2\ln(2/\delta)}{2n}}. \tag{3.4}$$

Chernoff and Hoeffding bounds are independent from the distribution generating examples. They are applicable in all situations where observations are independent and generated by a stationary distribution. Due to their generality they are conservative, that is they require more observations than when using distribution dependent bounds.

The Chernoff bound is multiplicative and its error is expressed as a relative approximation. The Hoeffding bound is additive and the error is absolute. While the Chernoff bound uses the sum of events and requires the expected value for the sum, the Hoeffding bound uses the expected value and the number of observations.

3.3.1 Maintaining Simple Statistics from Data Streams

The recursive version of the sample mean is well known:

$$\bar{x}_i = \frac{(i-1)\times\bar{x}_{i-1} + x_i}{i}. \tag{3.5}$$

In fact, to incrementally compute the mean of a variable, we need only to maintain in memory the number of observations (i) and the sum of the values seen so far $\sum x_i$. Some simple mathematics allow us to define an incremental version of the standard deviation. In that case we need to store three quantities: the number of data points i; the sum of the i points $\sum x_i$; and the sum of the squares of the i data points $\sum x_i^2$

$$\sigma_i = \sqrt{\left(\sum x_i^2 - \left(\sum x_i\right)^2 / i\right) / (i-1)}. \tag{3.6}$$

Another useful measure that can be recursively computed is the correlation coefficient. Given two streams x and y, we need to maintain the sum of each stream ($\sum x_i$ and $\sum y_i$), the sum of the squared values ($\sum x_i^2$ and $\sum y_i^2$), and the sum of the cross-product ($\sum(x_i \times y_i)$). The exact correlation is:

$$\operatorname{corr}(a,b) = \frac{\sum(x_i \times y_i) - \frac{\sum x_i \times \sum y_i}{n}}{\sqrt{\sum x_i^2 - \frac{(\sum x_i)^2}{n}}\sqrt{\sum y_i^2 - \frac{(\sum y_i)^2}{n}}}. \tag{3.7}$$

We have defined the *sufficient statistics* necessary to compute the mean, standard deviation, and correlation on a time series. The main interest of these formulas is that they allow us to maintain exact statistics (mean, standard deviation, and correlation) over an eventually infinite sequence of numbers without storing in memory all the numbers. Quantiles, median and other statistics [2] used in range queries can be computed from histograms, and other summaries that are discussed later in this chapter. Although these statistics are used everywhere they are of limited use in the stream problem setting. In most applications recent data is the most relevant one. To fulfill this goal, a popular approach consists of defining a time window covering the most recent data.

3.3.2 Time Windows

Time windows are a commonly used approach to solve queries in open-ended data streams. Instead of computing an answer over the whole data stream, the query (or operator) is computed, eventually several times, over a finite subset of tuples. In this model, a time stamp is associated with each tuple. The time stamp defines when a specific tuple is valid (e.g. inside the window) or not. Queries run over the tuples inside the window. However, in the case of joining multiple streams the semantics of time stamps is much less clear e.g. the time stamp of an output tuple.

Several window models have been used in the literature. The following are the most relevant.

3.3.2.1 Landmark Windows

Landmark windows [14] identified relevant points (the landmarks) in the data stream and the aggregate operator uses all record seen so far after the landmark. Successive windows share some initial points and are of growing size. In some applications, the landmarks have a natural semantic. For example, in daily basis aggregates the beginning of the day is a landmark.

3.3.2.2 Sliding Windows

Most of the time, we are not interested in computing statistics over all the past but only in the *recent* past. The simplest approach are *sliding windows* of fixed size. These type of windows are similar to *first in, first out* data structures. Whenever an element j is observed and inserted in the window, another element $j - w$, where w represents the window size, is forgotten.

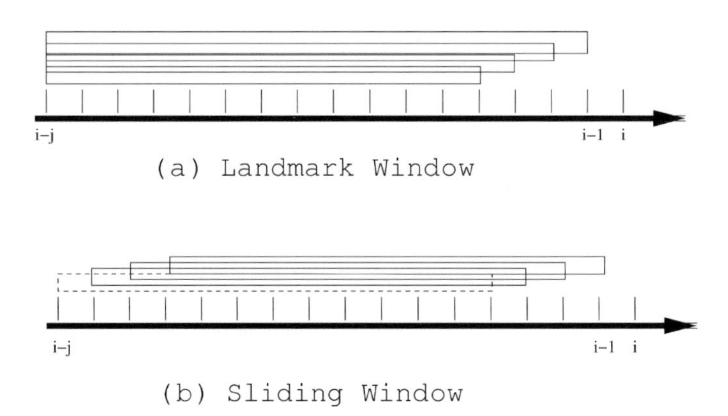

Fig. 3.1 Landmark and sliding windows

Suppose we want to maintain the standard deviation of the values of a data stream using only the last 100 examples, that is in a fixed time window of dimension 100. After seeing observation 1,000, the observations inside the time window are $x_{901}, \ldots, x_{1,000}$, and the sufficient statistics after seeing the 1,000th observation are $A = \sum_{i=901}^{1,000} x_i$; B=$\sum_{i=901}^{1,000} x_i^2$. Whenever the 1,001th value is observed, the time window moves 1 observation, forgets observation 901 and the updated sufficient statistics are: $A = A + x_{1,001} - x_{901}$ and $B = B + x_{1,001}^2 - x_{901}^2$. Due to the necessity to forget old observations, we need to maintain in memory all the observations inside the window. The same problem applies for time windows whose size changes with time. In the following sections we address the problem of maintaining approximate statistics over sliding windows in a stream, without storing in memory all the elements inside the window.

3.3.2.3 Tilted Windows

Previous windows models use a catastrophic forget; that is, any past observation either is in the window or it is not inside the window. In tilted windows, the time scale is compressed. The most recent data are stored inside the window at the finest detail (granularity). Oldest information is stored at a coarser detail, in an aggregated way. The level of granularity depends on the application. This window model is designated a *tilted time window*. Tilted time windows can be designed in several ways.

Han and Kamber [19] presented two possible variants: *natural tilted time windows*, and *logarithm tilted windows*. Illustrative examples are presented in Fig. 3.2. In the first case, data are stored with granularity according to a natural time taxonomy: last hour at a granularity of 15 minutes (4 points), last day in hours (24 points), last month in days (32 points) and last year in months (12 points). In the case of logarithmic tilted windows, given a maximum granularity with periods of t, the granularity decreases logarithmically as data are older. As time goes by, the window stores the last time period t, the one before that, and consecutive aggregates of less granularity (two periods, four periods, eight periods etc.).

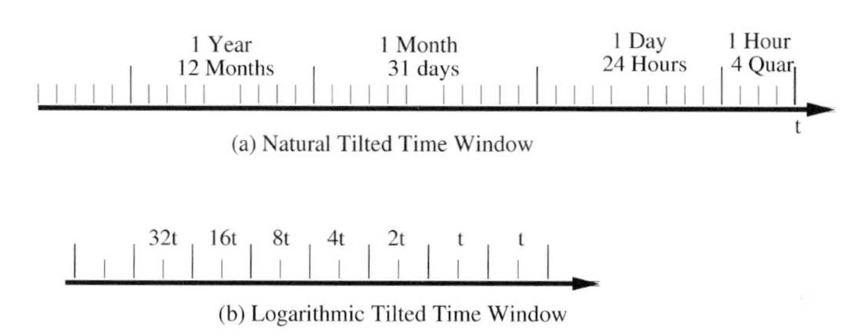

Fig. 3.2 Tilted Time Windows. The top figure presents a *natural tilted time window*, the figure in the bottom *logarithm tilted windows*

3.4 Basic Streaming Algorithms

With new data constantly arriving even as old data are being processed, the amount of computation time per data element must be low. Furthermore, since we are limited to a bounded amount of memory, it may not be possible to produce exact answers. High-quality approximate answers can be an acceptable solution. Two types of techniques can be used: data reduction and summaries. In both cases, queries are executed over a summary, a compact data-structure that captures the distribution of the data. In the latter approach, techniques are required for storing summaries or synopsis information about previously seen data. Certainly, large summaries provide more precise answers. So, there is a trade-off between the size of summaries and the overhead to update summaries and the ability to provide precise answers. In both cases, they must use data structures that can be maintained incrementally. The most commonly used techniques for data reduction involve: *sampling*, *synopsis* and *histograms*, and *wavelets*.

3.4.1 Sampling

Sampling involves loss of information: some tuples are selected for processing while others are skipped. Instead of dealing with an entire data stream, we can sample instances at periodic intervals. If the rate of arrival data in the stream is higher than the capacity to process data, sampling is used as a method *to slow down* data. Another advantage is the use of offline procedures to analyse data, eventually for different goals. Nevertheless, sampling must be done in a principled way in order to avoid missing relevant information.

The most time-consuming operation in database processing is the *join* operation. The join operation is a blocking operator, it pairs up tuples from two sources based on key values in some attribute(s). Sampling has been used to find approximate answers whenever a join between two (or more) streams is required. The main idea consists of executing a random sample from each input and then joining the samples [10]. The output will be much smaller than the join over the full streams. If the output is *representative* of the original, the approximate answers are representative of the original output. Traditional sampling algorithms require knowledge of the number of tuples. They are not applicable to the streaming setting. This way, join over streams requires sequential random sampling.

The *reservoir sampling* technique [28] is the classic algorithm to maintain an online random sample. The basic idea consists of maintaining a sample of size s, called the reservoir. As the stream flows, every new element has a certain probability of replacing an old element in the reservoir. Extensions to maintain a sample of size k over a count-based sliding window of the n most recent data items from data streams appear in [4]. Longbo et al. [22] presented a stratified multistage sampling algorithm for time-based sliding window (SMS Algorithm). The SMS algorithm takes a different sampling fraction in different strata from the time-based sliding window, and works even when the number of data items in the sliding window varies dynamically over time.

The literature contains a wealth of algorithms for computing quantiles [16] and distinct counts [15] using samples. Sampling is a general method for solving problems with huge amounts of data, and has been used in most streaming problems. Chaudhuri et al. [8] presented negative results for uniform sampling, mostly for queries involving joins. An active research area is the design of sampling-based algorithms that can produce approximate answers with a guarantee on the error-bound. As a matter of fact, sampling works by providing a compact description of much larger data. Alternative ways to obtain compact descriptions include *data synopsis*.

3.4.2 Data Synopsis

3.4.2.1 Frequency Moments

A useful mathematical tool employed to solve the above problems are the so-called *frequency moments*. A frequency moment is a number F^K, defined as $F^K = \sum_{i=1}^{v} m_i^K$, where v is the domain size, m_i is the frequency of i in the sequence, and $k >= 0$. F^0 is the number of distinct values in the sequence, F^1 is the length

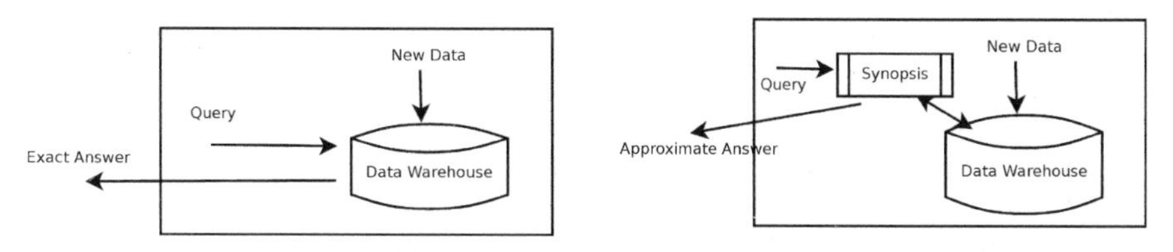

Fig. 3.3 Querying schema's: slow and exact answer vs fast but approximate answer using synopsis

of the sequence, F^2 is known as the self-join size (the repeat rate or Gini's index of homogeneity). The frequency moments provide useful information about the data and can be used in query optimizing. Alon et al. [1] showed that the second frequency moment of a stream of values can be approximated using logarithmic space. The authors also showed that no analogous result holds for higher frequency moments.

3.4.2.2 Hash Sketches

Hash functions are another powerful tool in stream processing. They are used to project attributes with huge domains into lower space dimensions. One of the earlier results is the Hash sketches (aka FM) for distinct-value counting introduced by Flajolet and Martin [13]. The basic assumption is the existence of a hash function $h(x)$ that maps incoming values $x \in [0, \ldots, N-1]$ uniformly across $[0, \ldots, 2^L - 1]$, where $L = O(\log N)$. Let $\text{lsb}(y)$ denote the position of the least-significant 1 bit in the binary representation of y. A value x is mapped to $\text{lsb}(h(x))$. The algorithm maintains a *Hash Sketch*, that is a bitmap vector of L bits, initialized to zero. For each incoming value x, set the $\text{lsb}(h(x))$ to 1. At each time-stamp t, let R denote the position of rightmost zero in the bitmap. R is an indicator of $\log(d)$, where d denotes the number of distinct values in the stream. Flajolet and Martin [13] proved that $E[R] = \log(\phi d)$, where $\phi = 0.7735$, so $d = 2^R/\phi$. This result is based on the uniformity of $h(x)$: $\text{Prob}[\text{BITMAP}[k] = 1] = \text{Prob}[10^k] = 1/2^{k+1}$. Assuming d distinct values, it is expected that $d/2$ to map to BITMAP[0], $d/4$ to map to BITMAP[1], $d/2^r$ map to BITMAP$[r-1]$.

The FM sketches are used for estimating the number of distinct items in a database (or stream) in one pass while using only a small amount of space.

3.4.3 Synopsis and Histograms

Synopsis and histograms are summarization techniques that can be used to approximate the frequency distribution of element values in a data stream. They are commonly used to capture attribute value distribution statistics for query optimizers (like range queries). A histogram is defined by a set of non-overlapping intervals. Each interval is defined by the boundaries and a frequency count. The reference technique is the V-Optimal histogram [18]. It defines intervals that minimize the frequency variance within each interval. Sketches, a special case of synopsis, provide probabilistic guarantees on the quality of the approximate answer (e.g. the answer is 10 ± 2 with probability 95%). Sketches have been used to solve the k-hot items [9].

3.4.3.1 Exponential Histograms

The *exponential histogram* is another histogram frequently used to solve counting problems. Consider a simplified data stream environment where each element comes from the same data source and is either 0 or 1. The goal consists of counting the number of 1's in a sliding window. The problem is trivial if the window can fit in memory. The problem we address here requires $O(\log(N))$ space, where N is the window size. Datar et al. [11] presented an exponential histogram strategy to solve this problem. The basic idea consists of using buckets of different sizes to hold the data. Each bucket has a time stamp associated with it. This time stamp is used to decide when the bucket is out of the window. Exponential histograms, other than buckets, use two additional variables, LAST and TOTAL. The variable LAST stores the size of the last bucket. The variable TOTAL keeps the total size of the buckets.

When a new data element arrives, we first check the value of the element. If the new data element is zero, ignore it. Otherwise create a new bucket of size 1 with the current time-stamp and incrementally advance the counter TOTAL. Given a parameter ϵ, if there are $|1/\epsilon|/2+2$ buckets of the same size, merge the two oldest buckets of the same size into a single bucket of double size. The largest time-stamp of the two buckets is used as the time-stamp of the newly created bucket. If the last bucket gets merged, we update the size of the merged bucket to the counter LAST.

Whenever we want to estimate the moving sum, we check if the oldest bucket is within the sliding window. If not, we drop that bucket, subtract its size from the variable TOTAL, and update the size of the current oldest bucket to the variable LAST. This procedure is repeated until all the buckets with the time stamps outside of the sliding window are dropped. The estimate of 1's in the sliding window is TOTAL-LAST/2.

The main property of the exponential histograms is that the size grows exponentially, i.e. $2^0, 2^1, 2^2, \ldots, 2^h$. To store N elements (one's) in the sliding window, only $O(\log(N)/\epsilon)$ are needed to maintain the moving sum and the error estimating. Datar et al. [11] proved that the error is bounded within a relative error ϵ.

3.4.4 Wavelets

Wavelet transforms are mathematical techniques in which signals are represented as a weighted sum of simpler, fixed building waveforms at different scales and positions. Wavelets attempt to capture trends in numerical functions, decomposing a signal into a set of coefficients. The decomposition does not preclude information loss, because it is possible to reconstruct the signal from the full set of coefficients. Nevertheless, it is possible to eliminate small coefficients from the wavelet transform introducing small errors when reconstructing the original signal. The reduced set of coefficients are of great interest for streaming algorithms.

A wavelet transform decomposes a signal into several groups of coefficients. Different coefficients contain information about characteristics of the signal at different scales. Coefficients at a coarse scale capture global trends of the signal, while coefficients at fine scales capture details and local characteristics.

The simplest and most common transformation is the Haar wavelet [21]. Any sequence $(x_0, x_1, \dots, x_{2n-1}, x_{2n})$ of even length is transformed into a sequence of two-component vectors $((s_0, d_0), \dots, (s_n, d_n))$. One stage of the Fast Haar-Wavelet Transform consists of:

$$\begin{bmatrix} s_i \\ d_i \end{bmatrix} = 1/2 \begin{bmatrix} 1 & 1 \\ 1 & -1 \end{bmatrix} \times \begin{bmatrix} x_i \\ x_{i+1} \end{bmatrix}.$$

The process continues, by separating the sequences s and d and transforming the sequence s.

As an illustrative example, consider the sequence (4, 5, 2, 1). Applying the Haar transform:

Resolution	Averages	Coefficients
4	(4, 5, 2, 1)	
2	(4.5, 1.5)	(−0.5, 0.5)
1	(3)	(1.5)

The output of the Haar transform is the sequence: (3, 1.5, −0.5, 0.5).

Wavelet analysis is popular in several streaming applications because most signals can be represented using a small set of coefficients. Matias et al. [23] presented an efficient algorithm based on multi-resolution wavelet decomposition for building histograms with application to databases problems, like selectivity estimation. Along the same lines, Chakrabarti et al. [7] proposed the use of *wavelet-coefficient synopses* as a general method to provide approximate answers to queries. The approach uses multi-dimensional wavelets synopsis from relational tables. Guha and Harb [17] proposed one-pass wavelet construction streaming algorithms with provable error guarantees for minimizing a variety of error-measures including all weighted and relative L_p norms.

3.5 Emerging Challenges and Future Issues

Data stream management systems development is an emerging topic in the computer science community, as they offer ways to process and query massive and continuous flows of data. Algorithms that process data streams deliver approximate solutions, providing a fast answer using few memory resources. In some applications, mostly data base oriented, an approximate answer should be within an admissible error margin. In general, as the range of the error decreases the space of computational resources goes up, but linear or sub-linear space data-structures, in the dimension of the inputs, are not applicable for data steams. The current trends in data streams involve research in algorithms and data structures with narrow error bounds. Other research issues involve the semantics of block-operators in the stream scenario, join over multi-streams and processing distributed streams with privacy preserving concerns.

Today, data are distributed in nature. Detailed data for almost any task are collected over a broad area, and stream in at a much greater rate than ever before. In particular, advances in miniaturization, the advent of widely available and cheap computer power, and the explosion of networks of all kinds led to life within inanimate things. Simple objects that surround us are gaining sensors, computational power, and actuators, and are changing from static, inanimate objects into adaptive, reactive systems. Sensor networks and social networks are present everywhere.

Acknowledgements This work was developed under project Adaptive Learning Systems II (POSI/EIA/ 55340/2004).

References

[1] N. Alon, Y. Matias, M. Szegedy, The space complexity of approximating the frequency moments. Journal of Computer and System Sciences, 58:137–147, 1999.

[2] A. Arasu, G.S. Manku, Approximate counts and quantiles over sliding windows. In: ACM Symposium on Principles of Database Systems (PODS), pp. 286–296. ACM Press, New York, 2004.

[3] B. Babcock, S. Babu, M. Datar, R. Motwani, J. Widom, Models and issues in data stream systems. In: P.G. Kolaitis (Ed.), Proceedings of the 21nd Symposium on Principles of Database Systems, pp. 1–16. ACM Press, New York, 2002.

[4] B. Babcock, M. Datar, Sampling from a moving window over streaming data. In: Proc. of the 13th Annual ACM SIAM Symposium on Discrete Algorithms, pp. 633–634. ACM/SIAM, New York/Philadelphia, 2002.

[5] C. Bettini, S.G. Jajodia, S.X. Wang, Time Granularities in Databases, Data Mining and Temporal Reasoning. Springer, Berlin, 2000.

[6] D. Carney, U. Çetintemel, M. Cherniack, C. Convey, S. Lee, G. Seidman, M. Stonebraker, N. Tatbul, S.B. Zdonik, Monitoring streams—a new class of data management applications. In: VLDB, pp. 215–226, 2002.

[7] K. Chakrabarti, M. Garofalakis, R. Rastogi, K. Shim, Approximate query processing using wavelets. VLDB Journal: Very Large Data Bases, 10(2–3):199–223, 2001.

[8] S. Chaudhuri, R. Motwani, V.R. Narasayya, On random sampling over joins. In: SIGMOD Conference, pp. 263–274, 1999.

[9] G. Cormode, S. Muthukrishnan, What's hot and what's not: tracking most frequent items dynamically. In: ACM Symposium on Principles of Database Systems (PODS), pp. 296–306, 2003.

[10] A. Das, J. Gehrke, M. Riedewald, Approximate join processing over data streams. In: Proc. of the ACM SIGMOD International Conference on Management of Data, pp. 69–84, 2003.

[11] M. Datar, A. Gionis, P. Indyk, R. Motwani, Maintaining stream statistics over sliding windows. In: Proceedings of 13th Annual ACM-SIAM Symposium on Discrete Algorithms, pp. 635–644. Society for Industrial and Applied Mathematics, 2002.
[12] M. Fang, N. Shivakumar, H. Garcia-Molina, R. Motwani, J.D. Ullman, Computing iceberg queries efficiently. In: Proc. 24th Int. Conf. Very Large Data Bases, VLDB, pp. 299–310, 1998.
[13] P. Flajolet, G.N. Martin, Probabilistic counting algorithms for data base applications. Journal of Computer and System Sciences, 31(2):182–209, 1985.
[14] J. Gehrke, F. Korn, D. Srivastava, On computing correlated aggregates over continual data streams. In: SIGMOD Conference, pp. 13–24. ACM Press, New York, 2001.
[15] P.B. Gibbons, Distinct sampling for highly-accurate answers to distinct values queries and event reports. Very Large Data Boses Journal, 541–550, 2001.
[16] M. Greenwald, S. Khanna, Space-efficient online computation of quantile summaries. In: SIGMOD Conference, pp. 58–66, 2001.
[17] S. Guha, B. Harb, Wavelet synopsis for data streams: minimizing non-euclidean error. In: Proceeding of the Eleventh ACM SIGKDD International Conference on Knowledge Discovery in Data Mining, pp. 88–97. ACM Press, New York, 2005.
[18] S. Guha, K. Shim, J. Woo, Rehist: relative error histogram construction algorithms. In: VLDB 04: Proceedings of the 30th International Conference on Very Large Data Bases, pp. 288–299. Morgan Kaufmann, San Mateo, 2004.
[19] J. Han, M. Kamber, Data Mining Concepts and Techniques. Morgan Kaufmann, San Mateo, 2006.
[20] C. Huitema, IPv6: The New Internet Protocol. Prentice Hall, New York, 1998.
[21] B. Jawerth, W. Sweldens, An overview of wavelet based multiresolution analyses. SIAM Rev., 36(3):377–412, 1994.
[22] Z. Longbo, L. Zhanhuai, Y. Min, W. Yong, J. Yun, Random sampling algorithms for sliding windows over data streams. In: Proceedings of the 11th Joint International Computer Conference—JICC. World Scientific, Singapore, 2005.
[23] Y. Matias, J.S. Vitter, M. Wang, Wavelet-based histograms for selectivity estimation, In: ACM SIGMOD International Conference on Management of Data, pp. 448–459, 1998.
[24] R. Motwani, P. Raghavan, Randomized Algorithms. Cambridge University Press, Cambridge, 1997.
[25] S. Muthukrishnan, Data streams: algorithms and applications. Now Publishers, 2005.
[26] V. Raman, B. Raman, J.M. Hellerstein, Online dynamic reordering for interactive data processing. In: The VLDB Journal, pp. 709–720, 1999.
[27] R. Snodgrass, I. Ahn, A taxonomy of time databases. In: SIGMOD '85: Proceedings of the ACM SIGMOD International Conference on Management of Data, pp. 236–246, USA. ACM Press, New York, 1985.
[28] J.S. Vitter, Random sampling with a reservoir. ACM Transactions on Mathematical Software, 11(1):37–57, 1985.

Chapter 4
Data Stream Processing in Sensor Networks

Mohamed Medhat Gaber

4.1 Introduction

A sensor network consists of small computational devices that are able to communicate over wireless connection channels [44,40]. Each of these computational devices is equipped with sensing, processing and communication facilities. The sensing facility can sense physical values about the environment such as temperature, humidity and light. The processing part is able to do computation on the sensed values and/or other received values from the neighbors. The communication part is able to listen and send to other sensor nodes.

The adoption of the sensor network technology will revolutionize our daily activities. Sensor networks will form a new world wide web that can read the physical world in real time [15]. Applications of sensor networks range from personal applications to scientific, industrial and business uses [11]. These applications will impact decision making at all levels, personally and professionally. Examples of these applications include habitat monitoring [31,39], animal control on a farm [10,38], traffic monitoring [19], underwater monitoring [41], fire detection [6], and smart homes [33].

Sensor networks generate data streams that need to be processed in real time for a wide range of applications. Therefore, there are two main aspects to this technology. The first is the scientific and technological aspect of sensor networks, which are discussed in Chap. 2. The other aspect is processing the resulting data streams, which is discussed in Chap. 3. This chapter is devoted to introducing data stream processing in sensor networks.

The distinction between traditional data stream processing and sensory data stream processing is important because sensory data streams have their own features. Elnahrawy [14] distinguished sensor streaming from traditional streaming in the following way:

M.M. Gaber
Tasmanian ICT Centre, CSIRO ICT Centre, GPO Box 1538, Hobart, TAS 7001, Australia
e-mail: Mohamed.Gaber@csiro.au

- The sensor data are a sample of the entire population. On the other hand, traditional streaming data such as web logs and stock market data streams represent the entire population of the data. Basically the sensor data depends on the sampling rate. Some applications require higher rate than the others. For example, a sensor network that measures the temperature for meteorological purposes could sample the data every five minutes. However, if we deploy the same sensor network for physical experimentation in a scientific laboratory, the sampling rate could be several times per second. Thus, the sampling rate is an application-dependent factor. The main theme in sensor networks is to convert continuous sensor readings to discrete ones.
- The sensor data streams are considered noisy by comparison with other traditional streaming data. The state-of-the-art sensing equipment onboard sensor nodes requires data verification and validation. The environmental effect on the deployed sensor networks can also play a negative role on the sensed values. For example, web logs and web click streams are considered accurate values compared with data generated from sensor networks. Data quality assurance and cleaning are important to ensure the reliability of sensor networks.
- The sizes of sensor data streams are usually less than traditional streaming data. This is valid for the current experimental deployment of wireless sensor networks. This will change in the near future with the expected large deployment of sensor networks to serve different applications.

These features combined with new research challenges have introduced a new field of study, data stream processing in sensor networks. This field is concerned with handling and processing sensed data streams in wireless sensor networks. We can broadly classify the processing tasks in this area into data management and data analysis/mining.

This chapter is organized as follows. Section 4.2 provides details about each class of data processing in sensor networks (data management and data mining). Section 4.3 highlights the research issues in data processing. Finally the chapter is summarized in Sect. 4.4.

4.2 Classification of Data Stream Processing in Sensor Networks

There are two classes of processing streaming data in sensor networks as pointed out in Sect. 4.1. This section provides a short introduction to each category of processing tasks.

4.2.1 Data Management

This class of processing tasks refers to handling, querying, scheduling, and storage of data streams generated in sensor networks. TinyDB [30], Cougar [42], and SS-

DQP [37] are among major data stream management systems developed for wireless sensor networks. Traditional data stream management systems include STREAM [2], Gigascope [22], and Aurora [1]. Traditional data stream management [3,34] and sensor network data stream management have several important differences.

- Spatial and temporal attributes play a major role in sensor networks. These attributes have no major effect on data streams in other applications such as telecommunication data streams and web logs and click streams. The need to know where and when the sensor reading has occurred is crucial to sensor network applications. For example, it is important to have exact information about the areas affected by a fire.
- Query design is based mostly on energy efficiency in sensor networks. Data streams generated from other sources have no energy constraints. For example, the number of phone calls in a city does not affect a query about phone calls in this particular city over the last hour. However, a query about temperature and humidity from a sensor network affects the energy consumed to sense these values, process them, and then send the results off.
- In-network data processing is the acceptable mode of operation in sensor networks. Data streams from other sources are processed outside of the network that produces them. Transferring all the sensed data streams to a central site to be processed is infeasible due to energy requirements and bandwidth limitations.
- Distributed processing of queries is the computational model in sensor networks. Centralized data stream processing is the common computational model in traditional data stream management systems.

4.2.2 Data Mining

Data stream mining in sensor networks is the second category of processing tasks. Clustering [36,4,26,13,25], outlier detection [7], classification and prediction [32,5], frequent pattern [27], time series [28] and change detection [21,35] have been studied for wireless sensor network applications.

Mining data streams has been an active area of research over the last few years. Sampling, load shedding and approximation techniques have been used in different ways to address the challenges of mining streaming data [17]. Data mining in sensor networks has brought new issues and challenges to the area. The computational power of sensor nodes is very limited. The state-of-the-art processors in sensor nodes lack the floating point operations. Emulation of these operations is a real burden on the sensor nodes' processor. There are several significant differences between traditional data stream mining and sensor network data stream mining.

- Data duplication in sensor networks is common due to similar environmental conditions over large areas. The data mining technique should take into consideration this data duplication. On the other hand, traditional mining of data streams has not dealt with the data duplication issue. For example, the variations in temperature and humidity are very limited even in large cities with exceptions of mountains and hills. Sensor networks deployed for this application read large amounts of duplicated data streams.

- Multi-level data mining in sensor network is essential. Data streams generated from individual nodes can build local models that are totally different than the global one. Models from individual nodes are required to be integrated in-network due to communication and energy limitations. Traditional data stream mining has not dealt with this problem.
- Real-time data cleaning is required to build an accurate model. Data streams generated from sensor networks are considered noisy, as discussed in Sect. 4.1. Since a wide range of sensor network applications are real time, there is a need for data cleaning in order to produce reliable data mining models.
- Adaptation to available resources is a key to the success of data mining in sensor networks. Sensor nodes have limited availability of computational resources. The data mining technique should be able to adapt to the current availability of resources [16]. Traditional data stream mining techniques have paid limited attention to this problem due to reasonable availability of computational resources to mine data streams in high-performance computational facilities. On the other hand and due to in-network processing requirements, it is a significant factor for data mining in sensor networks.

Having discussed the two classes of data stream processing in sensor networks, the following section discusses the research issues and challenges that face data management and mining in sensor networks.

4.3 Research Issues and Challenges

Data stream processing in wireless sensor networks faces a number of research issues and challenges [18,12]. Those issues include the following:

- *Energy efficiency* Sensor nodes operate on low-power batteries. Data management and mining techniques in sensor networks should be energy-aware. Sensor nodes consume a large percentage of their energy sending and/or receiving data streams [8]. Therefore, local processing of sensed data is crucial to sustaining operability. One of the common techniques used in data aggregation is to cluster sensor nodes, such as the HEED [43] and LEACH [20] protocols. Changing the sampling rate of the algorithm according to the availability of energy onboard the sensor node has also been investigated [36]. Context-awareness has been used to change the status of the sensor node from operation to sleep in order to conserve energy [9].
  The energy issue has been the main motivating research challenge in data processing in wireless sensor networks. Most of the proposed techniques in the literature have addressed this issue in one way or another.

- *Communication efficiency* Sensor networks operate over wireless connections with limited bandwidth. Large deployment of sensor networks represent a real challenge for data processing. Transferring data streams and/or output of the data processing technique is limited by the lack of bandwidth. Also, the connection can be unreliable. Data aggregation onboard sensor nodes is important to reduce the communication overhead. The use of data-reduction techniques such as Principal Component Analysis (PCA) has also been proposed [23].
  The communication issue is closely related to energy efficiency. Improving the efficiency of one implicitly improves the efficiency of the other.
- *Processing power* Sensor nodes have limited processing power to perform advanced computational tasks. Lack of floating point operations and limited processor speed have been an obstacle for large deployment of sensor networks. Approximation of floating point numbers and emulation of their operations have been the direct solution. Thus, hardware solutions are required for the success of sensor network applications.
  We note that most of the research studies done in this area have used simulation due to the limitations of state-of-the-art processors of sensor nodes.
- *Memory size* A wide range of traditional processing techniques require data to be resident in memory for processing. Limited availability of memory onboard sensor nodes makes these techniques unsuitable for sensor network applications. The use of data stream processing solves this issue.
- *Scalability* Real-deployed sensor networks will have thousands of nodes. Data processing techniques should be able to scale in order to process large amounts of distributed data streams. The scalability issue has been addressed [30].
- *Fault tolerance* Due to the large deployment of sensor networks to meet the application needs, fault tolerance is a crucial issue. Different fault tolerance techniques for wireless sensor networks can be found in [24].

The above research issues represent the driving force behind research in data stream processing in sensor networks.

4.4 Summary

This chapter introduces the area of data stream processing in wireless sensor networks. We have broadly classified the processing activities into two main categories: data management and data mining. Data management is concerned with querying and storage of data streams in sensor networks. Data mining is concerned with analysis of streaming data from sensor nodes. Research issues and challenges in both categories are discussed thoroughly. These issues include energy efficiency, communication efficiency, processing power, memory size, scalability and fault tolerance.

Part II discusses the data management techniques, and Part III addresses the data mining techniques. Finally, Part IV discusses the different applications of data processing in wireless sensor networks.

References

[1] D. Abadi, D. Carney, U. Cetintemel, M. Cherniack, C. Convey, C. Erwin, E. Galvez, M. Hatoun, J. Hwang, A. Maskey, A. Rasin, A. Singer, M. Stonebraker, N. Tatbul, Y. Xing, R. Yan, S. Zdonik, Aurora: a data stream management system (demonstration). In: Proceedings of the ACM SIGMOD International Conference on Management of Data (SIGMOD'03), San Diego, CA, June 2003.

[2] A. Arasu, B. Babcock, S. Babu, M. Datar, K. Ito, I. Nishizawa, J. Rosenstein, J. Widom, STREAM: The Stanford stream data manager demonstration description—short overview of system status and plans. In: Proceedings of the ACM Intl Conf. on Management of Data (SIGMOD 2003), June 2003.

[3] B. Babcock, S. Babu, M. Datar, R. Motwani, J. Widom, Models and Issues in Data Stream Systems. In: Proceedings of Principles of Database Systems (PODS'02), 2002.

[4] S. Bandyopadhyay, C. Gianella, U. Maulik, H. Kargupta, K. Liu, S. Datta, Clustering distributed data streams in peer-to-peer environments. Information Sciences, 176(14):1952–1985, 2006.

[5] G. Bontempi, Y.A. Borgne, An adaptive modular approach to the mining of sensor network data. In: Proceedings of 1st International Workshop on Data Mining in Sensor Networks as part of the SIAM International Conference on Data Mining, pp. 3–9, 2005.

[6] G. Boone, Reality mining: browsing reality with sensor networks. In: Sensors Online, vol. 21, September 2004.

[7] J. Branch, B. Szymanski, R. Wolff, C. Gianella, H. Kargupta, In-network outlier detection in wireless sensor networks. In: Proceedings of the 26th International Conference on Distributed Computing Systems (ICDCS), 2006.

[8] V. Cantoni, L. Lombardi, P. Lombardi, Challenges for data mining in distributed sensor networks. ICPR (1), 2006:1000–1007, 2006.

[9] S.K. Chong, S. Krishnaswamy, S.W. Loke, A context-aware approach to conserving energy in wireless sensor networks. PerCom Workshops, 2005:401–405, 2005.

[10] P. Corke, P. Sikka, P. Valencia, C. Crossman, G. Bishop-Hurley, D. Swain, Wireless ad hoc sensor and actuator networks on the farm. In: Fifth International Conference on Information Processing in Sensor Networks (IPSN), 2006.

[11] D. Culler, D. Estrin, M. Srivastava, Overview of sensor networks. IEEE Computer. 37(8) (2004), 41–49. Special Issue in Sensor Networks.

[12] S. Datta, K. Bhaduri, C. Giannella, R. Wolff, H. Kargupta, Distributed data mining in peer-to-peer networks. IEEE Internet Computing, 10(4):18–26, 2006. Invited Submission for Special issue on Distributed Data Mining.

[13] I. Davidson, A. Ravi, Distributed pre-processing of data on networks of Berkeley motes using non-parametric EM. In: Proceedings of 1st International Workshop on Data Mining in Sensor Networks as Part of the SIAM International Conference on Data Mining, pp. 17–27, 2005.

[14] E. Elnahrawy, Research directions in sensor data streams: solutions and challenges. In: DCIS Technical Report DCIS-TR-527, Rutgers University, 2003.

[15] J. Elson, D. Estrin, Sensor networks: a bridge to the physical world. In: Znati, Radhavendra, Sivalingam (Eds.), Chapter in Wireless Sensor Networks. Kluwer Academic, Dordrecht, 2004.

[16] M.M. Gaber, S. Krishnaswamy, A. Zaslavsky, On-board mining of data streams in sensor networks, In: S. Badhyopadhyay, U. Maulik, L. Holder, D. Cook (Eds.), A book chapter in Advanced Methods of Knowledge Discovery from Complex Data. Springer, Berlin, 2005.

[17] M.M. Gaber, A. Zaslavsky, S. Krishnaswamy, Mining data streams: a review. ACM SIGMOD Record, 34(2), ISSN: 0163-5808, 2005.

[18] J. Gehrke, S. Madden, Query processing in sensor networks. IEEE Pervasive Computing, 3(1):46–55, 2004.

[19] A. Guitton, A. Skordylis, N. Trigoni, Utilizing correlations to compress time-series in traffic monitoring sensor networks. In: IEEE Wireless Communications and Networking Conference (WCNC), 2007.
[20] W. Heinzelman, A. Chandrakasan, H. Balakrishnan, An application-specific protocol architecture for wireless microsensor networks. IEEE Transactions on Wireless Communications, 1(4):660–670, 2002.
[21] K. Hiramatsu, T. Hattori, T. Yamada, T. Okadome, Finding small changes using sensor networks. In: Proceedings of Ubicomp 2005 Workshop on Smart Object Systems, pp. 37–44, Tokyo, Japan, 2005.
[22] T. Johnson, C. Cranor, O. Spatscheck, Gigascope: a stream database for network application. In: Proceedings of the 2003 ACM SIGMOD International Conference on Management of Data, pp. 647–651, 2003.
[23] H. Kargupta, V. Puttagunta, M. Klein, K. Sarkar, On-board vehicle data stream monitoring using minefleet and fast resource constrained monitoring of correlation matrices. Next Generation Computing, 25(1):5–32, 2007. Invited Submission for Special Issue on Learning from Data Streams.
[24] F. Koushanfar, M. Potkonjak, A. Sangiovanni-Vincentelli, Fault tolerance techniques for wireless ad hoc sensor networks. In: IEEE Sensors 2002, vol. 2, pp. 1491–1496. IEEE, New York, 2002.
[25] D. Krivitski, A. Schuster, R. Wolff, Local hill climbing in sensor networks. In: Proceedings of 1st International Workshop on Data Mining in Sensor Networks as part of the SIAM International Conference on Data Mining, pp. 38–47, 2005.
[26] A. Kulakov, D. Davcev, Data mining in wireless sensor networks based on artificial neural-networks algorithms. In: Proceedings of 1st International Workshop on Data Mining in Sensor Networks as Part of the SIAM International Conference on Data Mining, pp. 10–16, 2005.
[27] C. Leung, Q. Khan, B. Hao, Distributed mining of constrained patterns from wireless sensor data, WI-IATW 248–251. In: IEEE/WIC/ACM International Conference on Web Intelligence and Intelligent Agent Technology (WI-IAT 2006 Workshops)(WI-IATW'06), 2006.
[28] J. Lian, K. Naik, L. Chen, Y. Liu, G. Agnew, Gradient boundary detection for time series snapshot construction in sensor networks. In: IEEE Transactions on Parallel and Distributed Systems (TPDS), 2007.
[29] S. Madden, M. Franklin, J. Hellerstein, W. Hong, TinyDB: an acquisitional query processing system for sensor networks. In: ACM TODS, 2005.
[30] U. Madhow, On scalability in sensor networks. In: Proceedings of Workshop on Information Theory and its Applications, UCSD Campus, USA, 2006.
[31] A. Mainwaring, J. Polastre, R. Szewczyk, D. Culler, J. Anderson, Wireless sensor networks for habitat monitoring. In: WSNA'02, 2002.
[32] S. McConnell, D. Skillicorn, A distributed approach for prediction in sensor networks. In: Proceedings of 1st International Workshop on Data Mining in Sensor Networks as Part of the SIAM International Conference on Data Mining, pp. 28–37, 2005.
[33] S. Meyer, A. Rakotonirainy, A survey of research on context-aware homes, wearable, invisible. In: Context-Aware, Ambient, Pervasive and Ubiquitous Computing (WICAPUC), 2003.
[34] S. Muthukrishnan, Data streams: algorithms and applications. In: Proceedings of the Fourteenth Annual ACM-SIAM Symposium on Discrete Algorithms, 2003.
[35] T. Palpanas, D. Papadopoulos, V. Kalogeraki, D. Gunopulos, Distributed deviation detection in sensor networks. ACM SIGMOD Record, 32(4):77–82, 2003.
[36] N.D. Phung, M.M. Gaber, U. Roehm, Resource-aware online data mining in wireless sensor networks. In: The Proceedings of 2007 IEEE Symposium on Computational Intelligence and Data Mining. IEEE Press, 2007.
[37] B. Scholz, M.M. Gaber, T. Dawborn, R. Khoury, E. Tse, Efficient time triggered query processing in wireless sensor networks. In: Proceedings of the International Conference on Embedded Systems and Software ICESS-07 to be held in Daegu, Korea, 14–16 May. Springer, 2007.

[38] P. Sikka, P. Corke, L. Overs, Wireless sensor devices for animal tracking and control. In: Proceedings of 29th Conference on Local Computer Networks LCN2004. Tampa, Florida, 2004.
[39] R. Szewczyk, A. Mainwaring, J. Polastre, D. Culler, An analysis of a large scale habitat monitoring application. In: ACM SenSys, November 2004.
[40] M. Tubaishat, S. Madria, Sensor networks: an overview. IEEE Potentials, 22(2):20–23, 2003.
[41] I. Vasilescu, K. Kotay, D. Rus, M. Dunbabin, P. Corke, Data collection, storage and retrieval with an underwater sensor network. In: Proceedings of IEEE SenSys, pp. 154–165, 2005.
[42] Y. Yao, J. Gehrke, The cougar approach to in-network query processing in sensor networks. PACM SIGMOD Record, 31(3):9–18, 2002.
[43] O. Younis, S. Fahmy, Heed: a hybrid, energy-efficient, distributed clustering approach for ad-hoc sensor networks. IEEE Transactions on Mobile Computing, 3(4):366–379, 2004.
[44] F. Zhao, L. Guibas, Wireless Sensor Networks: An Information Processing Approach. Morgan Kaufmann, San Mateo, 2004.

Part II
Data Stream Management Techniques in Sensor Networks

Chapter 5
Data Stream Management Systems and Architectures

M.A. Hammad, T.M. Ghanem, W.G. Aref, A.K. Elmagarmid, and M.F. Mokbel

5.1 Introduction

A growing number of applications in several disciplines—sensor networks, Global Positioning Systems (GPSs), Internet, retail companies, and the phone network industry, for example—deal with a new and challenging form of data; namely data streams. In these applications, data evolve over time in an unpredictable and bursty arrival fashion, representing *streams* of sensor-measured values, locations of moving objects, network traffic, sales transactions or phone call records. In contrast to conventional data management applications, a major requirement of streaming applications is to continuously monitor and possibly react to new, interesting events at the input streams. Since input streams are unbounded, stream applications may specify that user queries should process only recent input data (e.g., data items that arrive within the last hour). This type of queries is termed time-based sliding window queries or SWQs for short.

This chapter describes our efforts to address the research challenges that arose while building Nile, a prototype data stream management system. Nile focuses on executing sliding-window queries over data streams and provides an operational platform that facilitates experimental data stream management research. Specifically, Nile includes novel technologies that address the following fundamental research challenges.

M.A. Hammad
University of Calgary, Calgary, AB, Canada
e-mail: hammad@ucalgary.ca

T.M. Ghanem · W.G. Aref · A.K. Elmagarmid
Purdue University, West Lafayette, IN, USA

M.F. Mokbel
University of Minnesota, Minneapolis, MN, USA

- *Correctness of SWQ.* SWQ introduces a new semantic. For example, in contrast to a SQL query that processes fixed multi-sets of input data and produces a fixed multi-set of output data. SWQ processes continuous flows of input data streams and produces a continuous flow of output data stream. Therefore, in order to judge new algorithms of evaluating SWQ, a correctness criterion must be in place. One approach to define a correct execution of SWQ is to map the outputs from SWQ at specific time instants to the outputs from repeatedly executing a SQL query with specific properties (described in Sect. 5.3). For now, we will refer to this SQL query as the time-instant, equivalent query or Q_c for short.
- *Efficient Execution of SWQ.* SWQs introduce significant challenges to conventional query processors for the following primary reason: with a high input rate, SWQ will encounter significant processing overhead to produce a new result as a new data item arrives. Nile adopts the correctness view of SWQ as repeated evaluations of Q_c. Therefore, Nile progressively builds on the inputs, intermediate data structures, and output of Q_c at one time instant to generate inputs, intermediate data structures, and output of Q_c at the following time instant. At the query level, Nile includes two approaches to progressively evaluate SWQs; namely the time probing approach and the negative tuple approach.

Moreover, the following research goals were addressed using Nile as the underlying system.

- *Evaluating Multi-way Stream Window Joins.* The multi-way join is a common operation in several streaming applications. When combined with the windowed execution, the continuous evaluation of the multi-way window join (or W-join for short) needs to instantly react to new tuples as well as expired tuples. The research in W-join uses the Nile progressive evaluation approach and features novel algorithms to maintain the join states (i.e., the hash tables maintained by the join operation).
- *Scalable Execution of Multiple SWQs.* The data stream processing architecture involves potentially large numbers of SWQs. The availability of a collection of standing queries in stream database systems raises the potential for aggressively sharing the processing required by multiple queries. Window joins are at the core of emerging architectures for continuous query processing over data streams. A naive approach would treat identical joins with different window constraints as having different signatures and would execute them separately. In this research we formulate the problem of shared window join processing and describe scheduling algorithms that prioritize such shared execution to reduce the average response time per query while preserving their original semantics.

This chapter is structured as follows. Section 5.2 presents the Nile representations of streams and SWQs. The SWQ correctness criterion is presented in Sect. 5.3. Section 5.4 describes the progressive evaluation of SWQ at the operator and the query-plan levels. In Sects. 5.5 and 5.6 we present the multi-way stream window join and the shared execution of multiple window joins over data streams, respectively. The related work is presented in Sect. 5.7 and we conclude the chapter with a summary in Sect. 5.8.

5.2 Nile System Design

5.2.1 The Stream Model

A stream is an infinite sequence of data items, where the items are added to the sequence over time and the sequence is ordered by the time stamp at which each item is added to the stream. Accordingly, we view each stream data item as a tuple with a schema $\langle V, TS \rangle$, where V is the set of attributes that describe the data item and TS is a system-generated time-stamp attribute. TS is the primary key of a stream and values of TS are monotonically increasing. Practically, the time-stamp values at one stream are independent from the time-stamp values of another stream. Therefore, Nile assumes no global ordering of the time-stamp values among all input streams.

5.2.2 System Architecture

Nile follows a centralized architecture in which data streams continuously arrive to be processed against a set of standing continuous queries. Nile is based on an open-source, object-relational DBMS, PREDATOR [14], which provides an environment to define new data types and an extensible query processing engine.

Figure 5.1 gives a system overview of the Nile stream data management system. The stream data sources provide infinite streams of raw data. The drivers receive raw streams from the sources and prepare them for query processing operations.

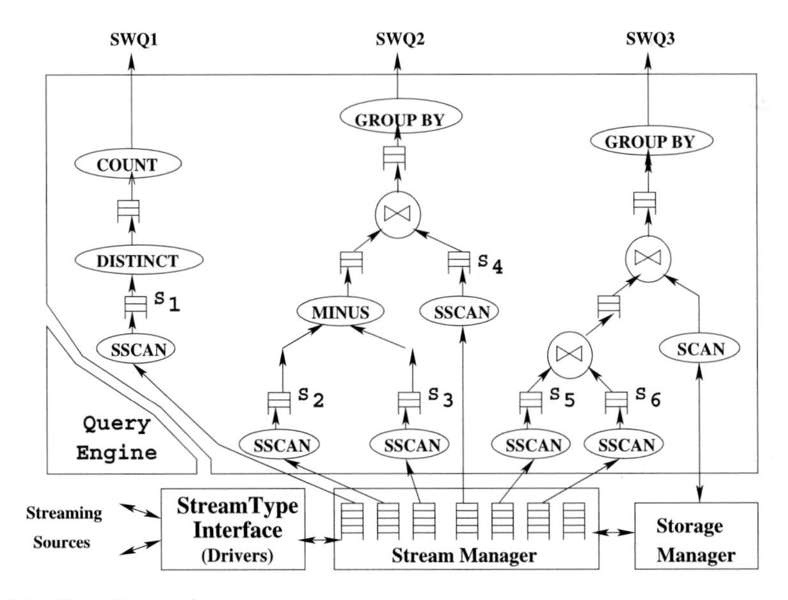

Fig. 5.1 Nile architectural components

The functionality of each driver depends on the specification of each source stream. For example, one driver would communicate over the network with a remote site that maintains the sales transactions of a group of WalMart stores. A second driver receives a stream of an MPEG video and prepares a frame-by-frame intermediate stream for processing by the query engine. Stream drivers are implemented in Nile as part of a user-defined data type. Nile includes a new component, which we refer to as *Stream Manager* that handles multiple incoming streams and acts as a buffer between the stream source and the stream query processor. The Stream Manager registers new stream-access requests, retrieves data from the registered streams into local stream buffers, and supplies data to the query processor. In the following we provide an overview of the stream representation as a data type, the interface between the stream data type and the query execution engine, and the window specification in continuous queries.

5.2.3 Stream Representation

Nile includes a new data type (SDT) to define the streaming attribute. A table with an SDT attribute is termed a *stream table*. During query execution, each access to the stream table returns a new tuple with a new SDT value. SDT is an object type that includes variables and routines. The variables store information such as the communication port to read stream values, the URL of the web page etc. The user must define one of the variables to store the run-time value of the SDT attribute and another variable to store the time-stamp value. Also, the user needs to provide the implementations of the following interface functions: *InitStream*, *ReadStream*, and *CloseStream*. These functions represent the basic routines that are called by other stream processing components of Nile. InitStream performs the necessary initializations, allocates resources and starts up communication with the source stream. A call to ReadStream returns a new run-time value from the source stream and assigns a new value of the time-stamp variable. The default time-stamp value is the time at which the run-time value was received by Nile, but the user may override the default with a different value within the ReadStream implementation code. In both cases Nile verifies that the time-stamps in each stream are monotonically increasing. CloseStream is invoked when no further data are needed from the source stream.

The Internal Representation of Stream Tuples. Nile uses an internal representation of the input tuples that has the schema $\langle V, TS_b, TS_e, \text{IsNegative}\rangle$, where V includes the SDT attribute, TS_b (*the begin time-stamp attribute*) stores the largest time-stamp at which the tuple becomes part of the stream and TS_e (*the end time-stamp attribute*) stores the earliest time-stamp at which the tuple is no longer part of the stream. The IsNegative attribute stores either FALSE for a *positive tuple* or TRUE for a *negative tuple*. Positive and negative tuples will be explained later in Sect. 5.4. For a base stream tuple (e.g., the tuples in Streams S_1 and S_2 of Fig. 5.1) with values v and a time-stamp ts_b, the internal representation is $\langle v, t_b, t_b + w,$

FALSE⟩, where w is the stream window size. For an intermediate stream tuple (e.g., tuples in Streams S_4 and S_7 of Fig. 5.1), the values of TS_b and TS_e attributes are assigned by the operator as explained in Sect. 5.4.

5.2.4 Query Representation

Due to the unbounded nature of streams, queries over streams are often defined in terms of *windows* that limit the scope of interest over streaming data. For example, in modern data centers sensors are used to monitor the temperature and humidity at locations throughout the room. For a large data center, thousands of such sensors could be required. A control system monitors these sensors to detect possible cooling problems. One can model this example scenario as a system with two streams, one for temperature sensors and one for humidity sensors. The schema of the streams can be of the form (LocationId, Value, TS_b), where LocationId indicates a unique location in the data center, Value is the sensor reading, and TS_b is as described earlier. A sliding window query, SWQ_1, that continuously monitors the count of locations that have both humidity and temperature values above specific thresholds within a one-minute interval could be specified as follows:

```
SWQ1:
SELECT COUNT(DISTINCT A.LocationId)
FROM Temperature A, Humidity B
WHERE A.LocationId = B.LocationId and A.Value > Threshold_t
and B.Value > Threshold_h
WINDOW 1 min;
```

A second example query, SWQ_2, continuously reports the maximum temperature and humidity values per location in the last one hour interval as follows:

```
SWQ2:
SELECT A.LocationId, MAX(A.Value), MAX(B.Value)
FROM Temperature A, Humidity B
WHERE A.LocationId = B.LocationId
GROUP BY A.LocationId
WINDOW 1 hour;
```

The WINDOW clause in the query syntax indicates that the user is interested in executing the queries over the sensor readings that arrive during the time period beginning at a specified time in the past and ending at the current time. When such a query is run in a continuous fashion, the result is a sliding window query. Other than the WINDOW clause, the syntax of SWQ matches that of a conventional SQL query.

Window queries may have forms other than the time sliding window described in the preceding examples. One variation of the window query is to identify the window in terms of the number of tuples instead of the time units. Another variation is to define the beginning of the window to be a fixed rather than a sliding time.

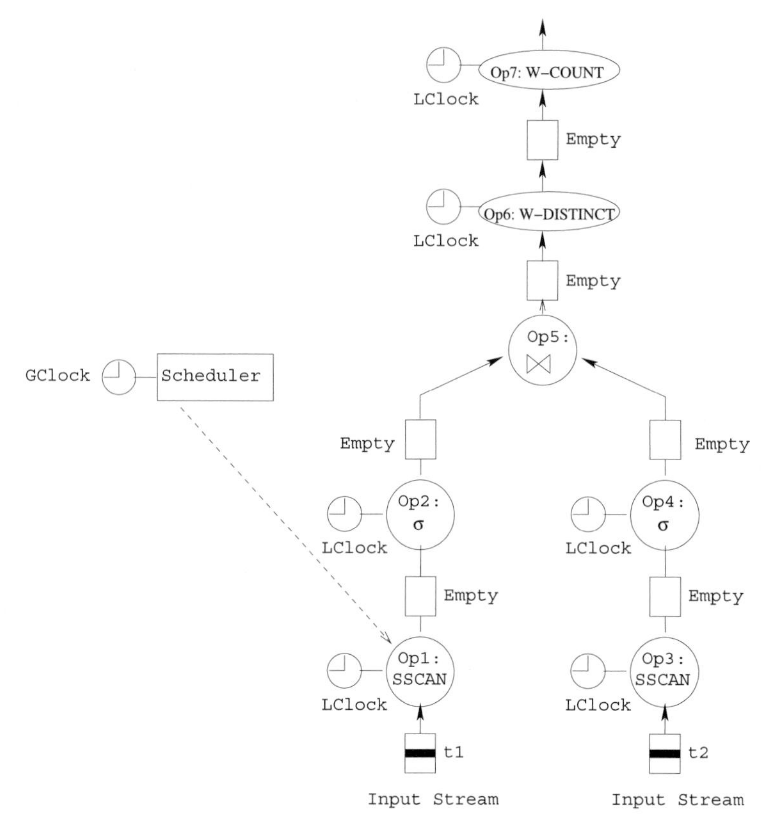

Fig. 5.2 A query execution plan in Nile

Other variations associate different windows with each stream [12] or with each pair of streams in a multi-way join [9].

The Query Execution Plan. A single SWQ consists of multiple operators organized in a query execution plan (QP). The inputs of QP can be either streams (*base streams*), relations (*base relations*) or both. Nile adopts the push-based pipelined query execution model in which query operators are independent execution units scheduled by a scheduler [8]. The intermediate streams and relations are buffered in First-In-First-Out (FIFO) queues while moving from a child operator to a parent operator in QP. Figure 5.2 gives the query execution plan of SWQ1.

5.2.5 The Stream Query Interface

Nile introduces the new *SScan* query operator to access the content of the stream table. SScan provides the typical interface functions Open(), GetNext(), and

Close() [6]. SScan.Open() accesses the stream table to retrieve the stored tuple. Next, SScan.Open() registers the SDT value with the Stream Manager (Fig. 5.3) for subsequent streaming. To be integrated in a push-based pipelined execution, SScan. Open() forks a process that repeatedly probes the Stream Manager for new data. Meanwhile, the Stream Manager retrieves data from the input stream sources into local buffers, which are accessed by the query execution engine.

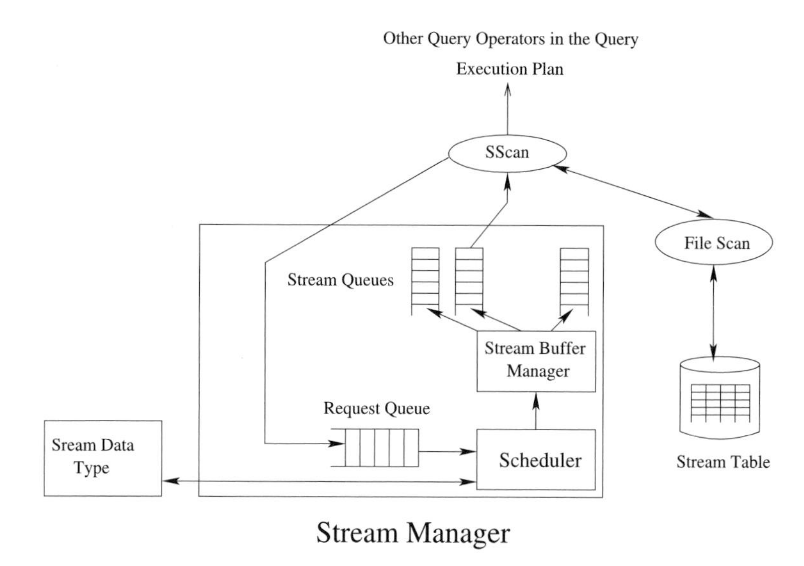

Fig. 5.3 The stream manager architecture

5.3 The Correctness Measure

Consider the following question: for a stream application, say A, that executes SWQ, *what should A expect as an output from* SWQ *at a time instant* T*?* An intuitive answer to the above question is that, at time T, A expects to receive the multi-set of output tuples that result from applying the operators of SWQ over the multi-sets of input tuples at time T, which are defined as follows. Input tuples at time T refer to tuples stored in the input relations or tuples in the input streams with time stamps within the windows defined over the input streams. Concisely, Nile adopts the following correctness measure.

Let SWQ$(I_1, I_2, \ldots, I_n, w_1, w_2, \ldots, w_n)$ be a sliding window query, where I_j is the jth input data stream and w_j is I_j's sliding window. A correct execution of SWQ must provide, for each time instant T, output that is equivalent to that of a snapshot query Q_c that has the following properties: (1) Q_c is a relational query consisting of SWQ with its window clauses removed. (2) The input to Q_c from each stream I_j is the set of input tuples with time stamps between $T - |w_j|$ and T.

5.4 The Progressive Evaluation of Sliding-Window Queries

Current data stream management systems adopt one of two approaches to evaluate SWQ. First, the whole evaluation approach, which is adopted by systems such as Borealis [1] and by the research in punctuated streams [17]. Second, the progressive evaluation approach, which is adopted by Nile, STREAM [3], and TelegraphCQ [5]. In the whole evaluation approach and at each time instant the SWQ is evaluated from scratch considering the input tuples that lie within windows defined over input streams and that are stored in base tables. In this approach the state of SWQ at time instant T (i.e., the input, intermediate, and output streams, and the operators' states) is not shared with the state of SWQ at time instant $T + 1$. This approach smoothly integrates with conventional query processors; however, the performance suffers as SWQ processes input streams with high arrival rates. On the other hand, the progressive evaluation approach processes only changes between the state of SWQ at time instant T and the state of SWQ at time instant $T + 1$. Because of the state sharing, the progressive evaluation approach is expected to support higher stream input rates than that supported by the whole evaluation approach. However, the progressive evaluation approach challenges the conventional query pipeline execution as we illustrate in the following paragraphs.

Challenge 1: Incorrect Results. Figure 5.4a gives the SQL representation and the pipelined execution plan of SWQ_3 *"Continuously report the total sales of items with price greater than 4 in the last hour"*. The SELECT and SUM operators are scheduled using the conventional pipelined execution, which we will refer to as the *Input Triggered* approach. In the *Input Triggered* approach an operator is scheduled only when an input tuple exists at its input. $S1$, $S2$, and $S3$ represent the input stream, the output stream after the SELECT operator, and the final output stream after applying the SUM operator, respectively. Stream C represents the expected correct output from SWQ_3 when the query reacts to the arrival of new input as well as the expiration of the tuples exiting from the sliding window. For simplicity, in the example we assume that tuples arrive at equal intervals. We present the streams at times T_1, T_2, and T_5 in parts (I), (II), and (III) of Fig. 5.4a, respectively. At $S3$, the reported value for the sum is correct at times T_1 (28) and T_5 (20), but is incorrect in between. For example, the correct output at time T_2 is 22 (due to the expiration of the old tuple 6). Similarly, at time $T_2 + 1$ (not shown in the figure), the correct SUM is 13 (due to the expiration of tuple 9). However, because of the Input Triggered scheduling, the SUM operator will not identify its expired tuples until receiving an input at time T_5. Note that the SUM operator could have reported the missing values (e.g., 22 and 13) at time T_5. In this case, the output in $S3$ at time T_5 will match the correct output. However, this is totally dependent on the pattern of input data and will include a significant delay. For example, in $S3$, if both 22 and 13 are released immediately before 20, the output delays for each is $T_5 - T_2$ and $T_5 - T_3$, respectively. Thus, at best, the Input Triggered approach would result in an *increased delay* of the output.

Challenge 2: No Support for Invalid Tuples. Figure 5.4b gives the SQL representation and the pipelined execution plan for SWQ_4 *"For each sold item in Sa-*

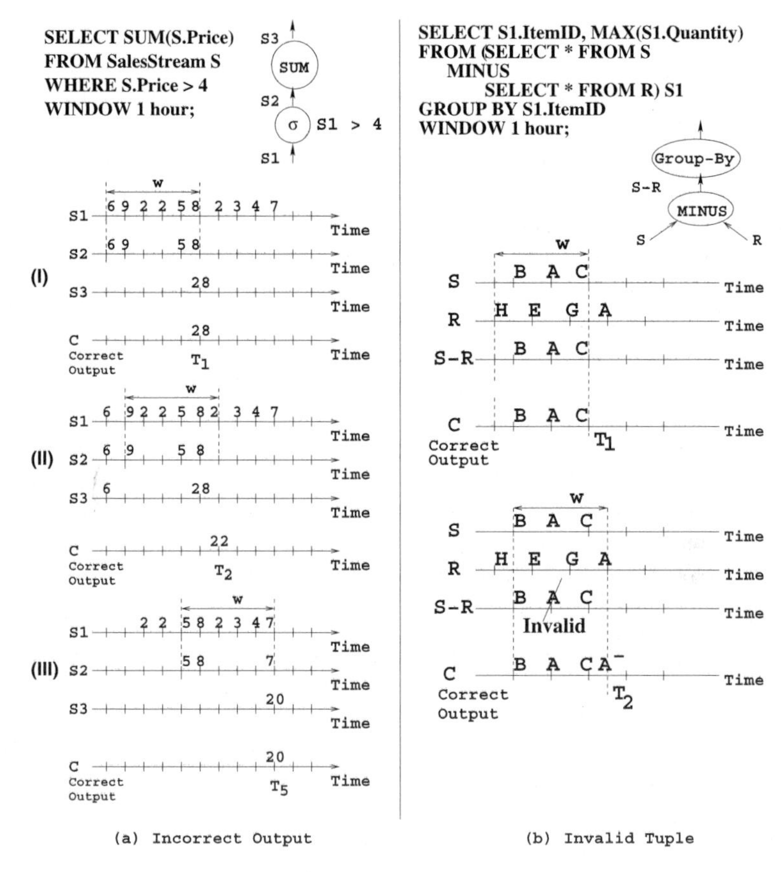

Fig. 5.4 Shortcomings of conventional pipelined execution

lesStream S and not in SalesStream R, continuously report the maximum sold quantity for the last hour". S and R represent the two input streams to the MINUS operator, while $S-R$ and C represent the output and the correct answer, respectively. Until time T_1, the MINUS operator provides a correct answer. However, at T_2, A is added to R and therefore A is no longer a valid output in $S-R$. Notice that A was still within the current window when it became *invalid*. In this case the MINUS operator needs to invalidate a previously reported output tuple. It can do so by generating an *invalid* output tuple. We represent the invalid output tuple as A^- in the correct output of Stream C at T_2. A^- removes any effect of the previously output A in Stream C. Note that, in this scheme, parent operators of the MINUS (e.g., Group-By in this case) must be able to react to the arrival of an *invalid* tuple. Thus, SWQ_4 indicates that the progressive evaluation of window operators needs to incorporate a new type of output/input tuples, i.e., *invalid tuples*.

In this following sections we present two approaches; namely the *Time Probing* and *Negative Tuple* approaches for correct and efficient execution of pipelined query plans. Both approaches require all tuples (both input and intermediate) to be time stamped. As discussed in Sect. 5.2.3 these time stamps are represented as *intervals* using two attributes associated with each tuple: TS_b and TS_e. For intermediate tuples that are created by the JOIN operators, values of TS_b and TS_e are assigned as follows: TS_b is set to the largest of the TS_b values of the tuples that contribute to the intermediate tuple and TS_e is set to the smallest of the TS_e values of those tuples. To maintain the notion of ordering during the pipeline execution, the query operators always produce their output tuples with TS_b monotonically increasing.

5.4.1 Approach I: Time Probing

Main idea. The *Time Probing* approach (TPA, for short) schedules a window query operator if a new tuple exists in the operator's input queue or if a stored tuple (e.g., a tuple in the hash table of a window join operator) expires. With TPA, a window operator expires an old tuple t when the operator guarantees that no tuple that arrives subsequently at the operator will have a TS_b value less than $t.TS_e$.

Implementation. Every operator in the pipeline stores the value of TS_b corresponding to the last processed tuple in a local variable termed *LocalClock*. Furthermore, each operator reports its LocalClock to the parent operator in the pipeline. Figure 5.5a illustrates TPA where each operator updates its LocalClock and propagates the LocalClock up in the pipeline. For simplicity, we omit the SScan operator from the query plan.

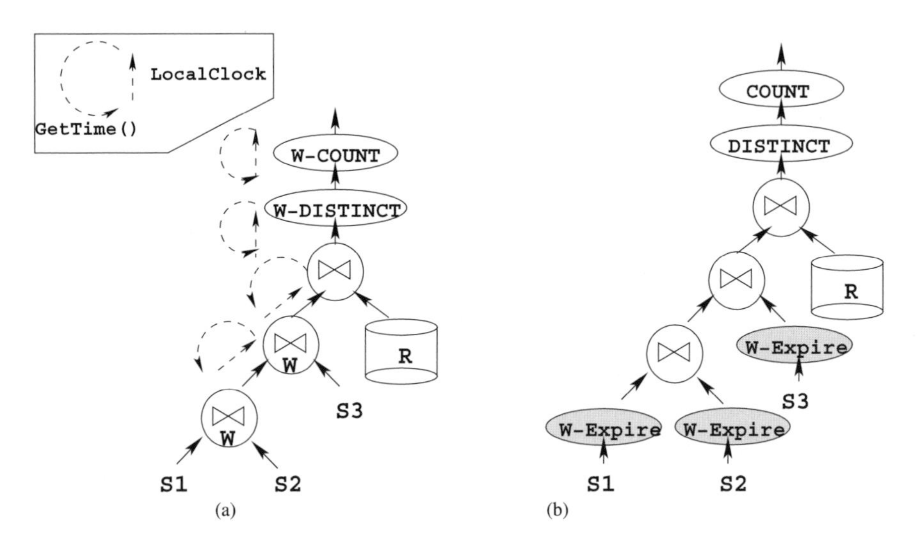

Fig. 5.5 A query plan with different scheduling approaches: **a** the time probing approach with window operators, **b** the negative tuple approach with no window operators

Example. Figure 5.6b gives the execution of TPA for the example given in Fig. 5.4a. For illustration, we report this example in Fig. 5.6a. Since the example assumes that input tuples exist at every clock tick, the SELECT operator can always update its LocalClock without further probing. At time T_2, the SUM operator did not receive any new tuples. Thus, it probes the SELECT operator asking for its local clock. The SELECT operator replies back with the time stamp of the tuple of value 2. Once the SELECT operator replies back, the SUM operator recognizes that it has to expire the old tuple with value 6. Thus, updating the answer of SWQ_3 to be 22. Similarly, the answer at T_3 is updated to be 13.

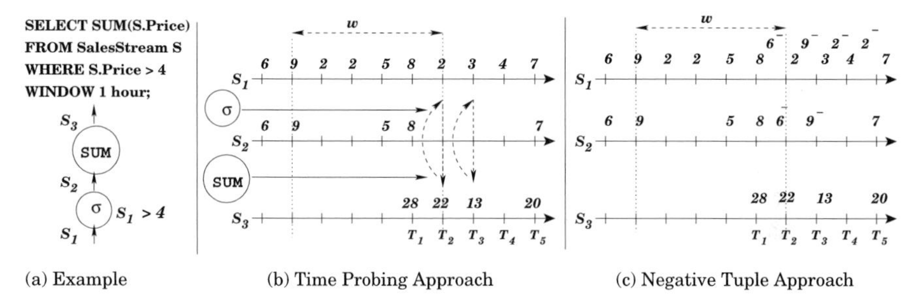

Fig. 5.6 Example of correct pipelined query execution

5.4.2 Approach II: Negative Tuple

Main idea. The *Negative Tuple* approach is inspired by the fact that, in general, window operators need to process *invalid tuples* (e.g., see Fig. 5.4b). A tuple t that is expired from a sliding window w can be viewed as a *negative tuple* t^- that goes through the pipeline following the footsteps of t. t^- cancels the effect of t in all query operators. *Negative tuples* are another form of *invalid tuples* that are produced by the MINUS operator. The *Negative Tuple* approach unifies the handling of both *invalid* and *negative* tuples. Query operators in the *Negative Tuple* approach are scheduled using the *Input Triggered* scheduling, i.e., an operator is scheduled only when a tuple exists at its input queue.

Implementation. Since *negative tuples* are synthetic tuples, we add a new leaf operator, *W-Expire*, that generates *negative tuples* for each input stream (see Fig. 5.5b). For any incoming tuple t, the *W-Expire* operator performs the following steps: (1) it stores t in its window structure; (2) it forwards t to the parent operator; and (3) it produces the *negative tuple* t^- when t is expired due to the sliding window. t^- has the same attributes as t and is tagged with a special *flag* that indicates that it is negative. Other query operators need to be equipped with special algorithms that process the *negative tuples*.

Example. Figure 5.6c gives the execution of the *Negative Tuple* approach for the example given in Fig. 5.6a. At time T_2, the tuple with value 6 expires. Thus, it appears in S_1 as a new tuple with value 6^-. The tuple 6^- passes the selection filter as it follows the footsteps of tuple 6. The SUM operator at time T_2 recognizes that it receives a *negative* input with value 6. Thus, it updates its output value to 22. Similarly at T_3, the SUM operator receives a *negative* tuple with value 9. Thus the result is updated to be 13. At T_5, the SUM operator receives a *positive* tuple with value 7. Thus, the output is updated to be 20.

A major advantage of the *Negative Tuple* approach is that it is very simple to implement. In fact, it almost incurs no coding overhead other than having the *W-Expire* operator. Adapting other query operators to handle *negative tuples* properly is already needed to support the MINUS operator. Such adaptation is needed even in the *Time Probing* approach. The simplicity of the *Negative Tuple* approach makes it suitable to extend existing database management systems to support unbounded data streams. Actual implementation of the *W-Expire* operator is placed within the SScan operator. Thus, *W-Expire* operators do not incur any additional scheduling overhead.

5.4.3 Stream-In Stream-Out Window Operators

In [7] the authors provided semantic descriptions of various window operators. In the following we present a classification of the window operators based on the positive/negative tuple paradigm. We use the term t^+ (*positive tuple*) to refer to a new tuple in the stream and the term t^- (*negative tuple*) to refer to an expired or invalid tuple in the stream. Based on the type of input and output tuples, we distinguish among four cases of window query operators (Fig. 5.7):

Case 1: A *positive* tuple, t^+_{out}, is produced at the output stream as a result of a *positive tuple*, t^+_{in}, being added to the input stream.

Case 2: A *negative* tuple, t^-_{out}, is produced at the output stream as a result of a *positive* tuple, t^+_{in}, being added to the input stream.

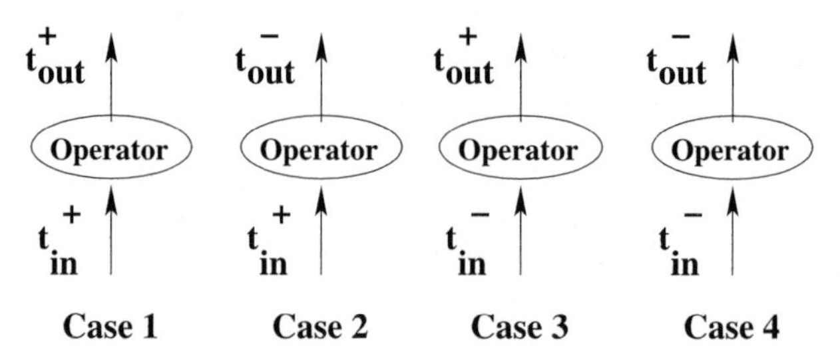

Fig. 5.7 Different relationships between tuples in input and output streams

Case 3: A *positive* tuple, t^+_{out}, is produced at the output stream as a result of a *negative* tuple, t^-_{in}, being added to the input stream.

Case 4: A *negative* tuple, t^-_{out}, is produced at the output stream as a result of a *negative* tuple, t^-_{in}, being added to the input stream.

Cases 1 and 4 can arise in all window operators. Consider for example the window join operator (W-join). For Case 1, when W-join receives an input tuple t^+_{in}, a new tuple t^+_{out} could be produced. For Case 4, when a negative tuple t^-_{in} becomes an input to the operator, and assuming that t^+_{in} results in an earlier output, then the joined tuple t^-_{out} should be produced. t^-_{out} indicates that the corresponding joined tuple is no longer part of the output stream. Cases 2 and 3 are special to some window operators such as the MINUS operator.

5.5 Extensions

As outlined in Sect. 5.1, Nile provides an integrated platform to experiment research ideas in data stream management. This section presents research contributions in data stream management made using Nile.

5.5.1 A Multi-Way Stream Window Join

A window join (W-join) is a class of join algorithms for joining multiple infinite data streams. W-join addresses the infinite nature of the data streams by joining stream data items that lie within a sliding window and that match a certain join condition. In addition to its general applicability in stream query processing, W-join can be used to track the motion of a moving object or detect the propagation of clouds of hazardous material or pollution spills over time in a sensor network environment. Consider the following three applications:

Application 1: Spotting the spread area of a pollution cloud using a sensor network. A user issues a query to monitor the propagation and expansion of a pollution cloud in a sensor network. The user specifies a time window for the propagation to occur based on wind speed and the maximum time for the cloud to propagate between two sensors and hence represents a window of interest over the readings between any two sensors.

Application 2: Tracking objects that appear in video data streams from multiple surveillance cameras. The objects are identified in each data stream and the maximum time for the object to travel through the monitoring devices defines an implicit time window for the join operation.

Application 3: Monitoring the sales from various department stores. A sales manager wants to find *common* items sold by all stores over *a sliding window of one hour*. Current transactions from each store represent streams of data containing information about items as they are sold. The execution requirements in these applications are different from those provided by conventional database systems due to the notion of sliding window. Therefore, both tuple addition and expiration must be considered during the monitoring process.

Each of these applications requires a join operation among data items in their input streams. The values of the window constraints in Applications 1 and 2 are derived from the applications' semantics (i.e., the maximum elapsed time to detect a gas or locate a moving object in two different data streams). However, in Application 3, the value of the window constraint is explicitly determined by the user. In the following we introduce the various forms of W-join and provide examples (expressed in a SQL-like representation) of each W-join form. In the SQL representation we use the WINDOW(A, B) to define the time window constraint among tuples in the streams A and B.

5.5.1.1 The Forms of Window Join

Form 1: *A binary window join* joins two input streams with a single window constraint. For example, the SQL query corresponding to Application 1 is (w is the maximum time needed for objects to move from Sensor A to Sensor B):

```
SELECT A.Gas
FROM Sensor1 A, Sensor2 B
WHERE A.GasId = B.GasId
WINDOW(A,B) = w
```

Form 2: *A path window join* joins multiple streams. In this case the window constraints connect the streams along one path. In Application 2 an object, Obj, needs different times ($w_1, w_2, \ldots$) to travel from one camera to the other. The user may issue the query:

```
SELECT A.Obj
FROM Camera1 A, Camera2 B, Camera3 C
WHERE similar(A.Obj,B.Obj) AND similar(B.Obj,C.Obj)
WINDOW (A,B)=w1 AND WINDOW (B,C)=w2
```

The *similar*() user-defined function determines when two objects that are captured by various cameras are similar.

Form 3: *A graph window join* joins multiple streams. In this case, the window constraints among the streams form a graph structure. For example, the following query tracks the spread of a hazardous gas that is detected by multiple sensors. The maximum times for the gas to travel through the sensors define the time windows w_1, w_2, w_3, and w_4:

```
SELECT A.Gas
FROM Sensor1 A, Sensor2 B, Sensor3 C, Sensor4 D
WHERE A.Gas=B.Gas AND B.Gas= C.Gas AND C.Gas= D.Gas
WINDOW(A,B)=w1 AND WINDOW(B,C)=w2 AND WINDOW(B,D)=w3
AND WINDOW(C,D)=w4
```

Notice that the window constraints may not exist among all possible pairs of streams (e.g., streams A and C may not be constrained by a window due to the existence of a barrier that prevents the gas from spreading directly).

Form 4: *A clique window join* is a special case of Form 3, where the window constraints exist between every pair of the joined streams. In the previous example if there are no barriers between any pair of sensors, then there exists a window constraint between every pair of sensors that represents the maximum time for the gas to travel between them. We consider the *uniform clique window join* as a special case of a clique window join, where all the streams are joined together using a single window constraint (same as Form 4 except that the sizes of the windows between each pair of the streams are equal). For example, to monitor the sales from multiple department stores using a sliding time window, say w, a user may issue the query:

```
SELECT A.ItemName
FROM Store1 A, Store2 B, Store3 C, Store4 D
WHERE A.ItemNum=B.ItemNum AND B.ItemNum=C.ItemNum AND
C.ItemNum=D.ItemNum
WINDOW = w
```

5.5.1.2 The W-join Operation

Figure 5.8 illustrates a W-join operation among five data streams (A–E). The indices of each tuple represent time stamps. The tuples from a single data stream arrive in time order; however, there is no implicit time order between two tuples that belong to two different data streams. The black dots correspond to tuples from each stream that satisfy the WHERE clause. The window constraint implies that tuples join only when they are a window of each other. Thus, the tuple $\langle a_7, b_6, c_5, d_{12}, e_8\rangle$ is a candidate for W-join. On the other hand, the tuple $\langle a_7, b_6, c_5, d_{12}, e_3\rangle$ is not a candidate, since e_3 and d_{12} are more than window w away from each other. The W-join in Fig. 5.8 represents *a uniform clique W-join*. A graph model of W-join can be obtained by representing tuples from the streams as nodes in a graph, where edges correspond to the window constraint (e.g., tuples from streams A and B must be within window of time from each other). With this model, the uniform clique W-join represents a complete graph (Fig. 5.8a). The non-uniform clique, graph, and path W-joins are given in Fig. 5.8b.

It is evident from the W-join operation that we need an efficient approach for continuously verifying window constraints among the input streams and for updating the *join buffers* (intermediate structures that hold tuples from each stream during the join) to contain only eligible tuples. A brute-force approach to verify window constraints among streams requires verifying the window constraint between each pair of n streams, adding $\binom{n}{2}$ additional comparisons for each input tuple, i.e., $O(n^2)$ comparisons. In the following we present an efficient algorithms with $O(n)$ complexity. Other algorithms to evaluate W-joins can be found in [9].

We describe the algorithm using an example W-join of five streams, A, B, C, D, E, as shown in Fig. 5.9. In this example the five streams are joined together using

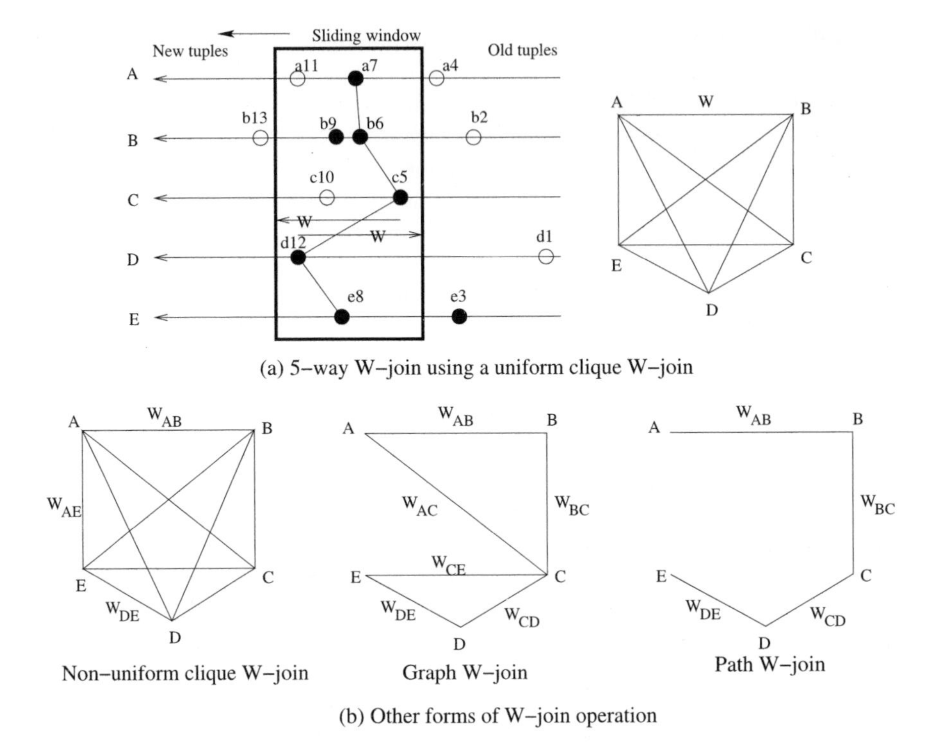

Fig. 5.8 Variations of W-join

a single window constraint of length w that applies between every two streams, i.e., a uniform clique W-join. For illustration of the algorithm, we assume that only the *black* dots (tuples) from each stream satisfy the join predicate in the WHERE clause (e.g., equality over objectID). We assume that the *vertical* bold arrow is currently pointing to the tuple about to be processed. For example, from the figure, Stream A is about to process the new tuple a_6. Figure 5.9 gives the positions of the tuples as they arrive over time; newer tuples are to the left of the stream and older tuples are to the right. There is no restriction that if the algorithm is processing a tuple from one stream then all tuples from the *other* streams that have earlier time-stamps must have been processed.

Figure 5.9a gives the state of the algorithm when processing tuple a_6 from Stream A and forming a window of length $2w$ centered at a_6. The algorithm iterates over all tuples of Stream B that are within window w of tuple a_6. These tuples are shown inside the interval over B. b_{12} is located within the window of a_6 (i.e., it is included in the interval) and satisfies the join predicate. The period is reduced in size to include a_6, b_{12}, and all tuples that are within w of a_6 and b_{12}. Notice that by shrinking the period, we maintain the number of necessary comparisons at two comparisons (at boundaries of the new shrunk period). This is in contrast to $O(N)$ comparisons if we verify the tuple inclusion using pairwise comparisons between the (new) coming

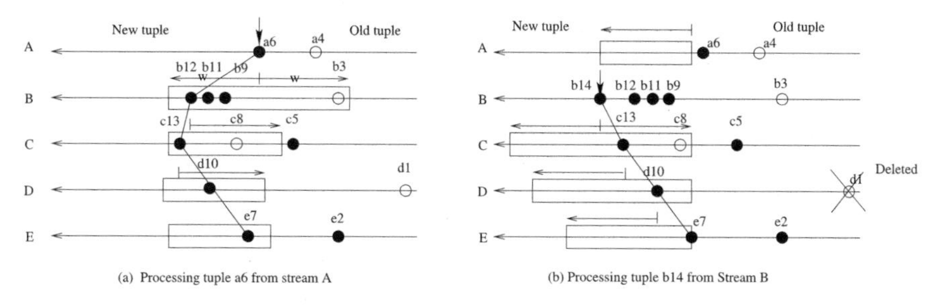

Fig. 5.9 Two iterations of ROW-join

tuple and each of the other tuples in the N data streams [2]. This new period is used to test tuples in Stream C, and is shown as an interval over Stream C in Fig. 5.9a. The process of checking the join condition is repeated for tuples in C. Since tuple c_{13} satisfies the join predicate and also lies inside the interval, a new period is calculated that includes tuples a_6, b_{12}, c_{13}, and all tuples that are within w of all of them. This period is shown as an interval over Stream D. In Stream D, d_{10} satisfies the join predicate and is located within the interval formed by a_6, b_{12}, and c_{13}. A new period is formed that includes the previous tuples and any further tuples within w of all of them. This period is shown as an interval over Stream E. The step is repeated for Stream E, and the 5_tuple, $\langle a_6, b_{12}, c_{13}, d_{10}, e_7 \rangle$ is reported as output. The algorithm recursively backtracks to consider other tuples in Streams D, then C and finally B. The final output 5_tuples in the iteration that start with tuple a_6 are: $\langle a_6, b_{12}, c_{13}, d_{10}, e_7 \rangle$, $\langle a_6, b_{11}, c_{13}, d_{10}, e_7 \rangle$, $\langle a_6, b_{11}, c_5, d_{10}, e_7 \rangle$, $\langle a_6, b_9, c_{13}, d_{10}, e_7 \rangle$, and $\langle a_6, b_9, c_5, d_{10}, e_7 \rangle$, respectively.

While iterating over Stream D, tuple d_1 is located at distance more than w from all the last tuples in Streams A, B, C, E, (i.e., tuples a_6, b_{12}, c_{13}, e_7). Tuple d_1 can be safely dropped from the join buffer of Stream D. After finishing with tuple a_6, the algorithm starts a new iteration using a different new tuple, if any. In the example of Fig. 5.9, we advance the pointer of Stream B to process tuple b_{14}. This iteration is shown in Fig. 5.9b, where periods over Streams C, D, E and A are constructed, respectively. This iteration produces no output, since no tuples join together in the constructed interval. The algorithm never produces spurious duplicate tuples, since in each iteration a new tuple is considered for the join (the newest tuple from a stream). The output tuples of this iteration must include the new tuple, thus duplicate tuples cannot be produced.

5.6 A Scalable Data Stream Management System

Consider the case of two or more queries, where each query is interested in the execution of a sliding window join over multiple data streams. We focus on concurrent queries with the same signature (e.g., queries that have the same join predicate over the same data streams, and where each query has a sliding window that represents its interest in the data). The goal is to share the execution of the different window joins to optimize the use of system resources. We illustrate this further using an example of two queries in Fig. 5.10 (Q_1 and Q_2 are SWQ$_1$ and SWQ$_2$ described in Sect. 5.2.4). In the figure, tuples arrive from the left and are tagged with their stream identifiers and time stamps. We indicate tuples that satisfy the join predicate (but not necessarily the window clause) by marking them with the same symbol (e.g., star, black circle etc.). In the figure, Q_1 performs a join between the two streams A and B, using predicate p with window size w_1 = one minute. Q_2 performs a join between the same two streams A and B, using predicate p with window size w_2 = one hour. There is an obvious overlap between the interests of both queries, namely, the answer of the join for Q_1 (the smaller window) is included in the answer of the join for Q_2 (the larger window). We refer to this as the *containment property* for the join operation; that is, the join answer of any query is also contained in the join answer of the queries having the same signature with larger windows.

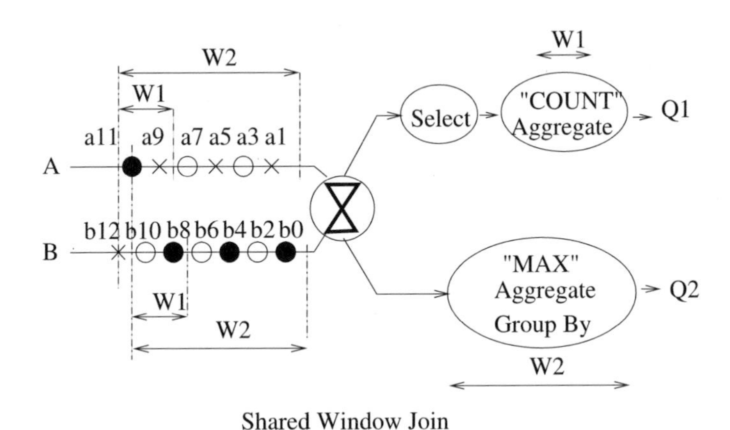

Fig. 5.10 The shared execution of two window joins

Executing both queries separately wastes system resources. The common join execution between the two queries will be repeated twice, increasing the amount of memory and CPU power required to process the queries. Implementing both queries in a single execution plan avoids such redundant processing.

The shared join produces multiple output data streams for each separate query. The output data streams are identified by their associated window sizes, and at least one query must be attached to each output data stream. The shared join operator is divided into two main parts: the join part and the routing part. The join part produces a single output stream for all queries and the routing part produces the appropriate output data streams for the various queries.

While shared execution has significant potential benefits in terms of scalability and performance, we need to ensure that such sharing does not negatively impact the behavior of individual queries. That is, the shared execution of multiple queries should be transparent to the queries. We define two objectives for such transparency:

1. The shared execution of window joins should abide by the isolated execution property, i.e., each window join, say j_w, that is participating in the shared execution, produces an output stream that is identical to the output stream that j_w produces when executing in isolation.
2. The response time penalty imposed on any query when a new query is included in a shared plan should be kept to a minimum.

In our example queries changing the order of the output during shared execution (a violation of objective 1 above) could potentially produce different COUNT and MAX results than isolated execution. In addition, when the shared execution imposes a high response time penalty for one query (e.g., Q_1), that query's output could be significantly delayed.

In [10] we describe and evaluate three scheduling algorithms that prioritize shared execution of window joins to reduce the average response time per query while preserving their original semantics. The first algorithm is termed Largest Window Only (LWO) and schedules the join operation to scan the largest window. LWO may penalize queries with smaller window sizes in terms of query response time. The second algorithm is Shortest Window First (SWF) that directly addresses the performance flaw identified for LWO by scheduling queries with small windows before queries with large windows. The third algorithm is Maximum Query Throughput (MQT) is motivated by the trade-offs between LWO and SWF and schedules an input tuple from one stream to join *with a partial window* over the second stream as long as the tuple will serve the maximum number of queries per unit time.

5.7 Related Work

The COUGAR [4] system and the work in [18] focuses on executing queries over sensor and stored data. Sensors are represented as a new data type, with special functions to extract sensor data when requested. COUGAR addresses scalability (increasing numbers of sensors) by introducing a virtual table where each row represents a specific sensor. The COUGAR system inspired many ideas in the early design phases of Nile. Specifically, the stream data type and the table representation of streams.

Seshadri et al. [15] presented the SEQ model and implementation for sequence databases. In this work a sequence is defined as a set with a mapping function to an ordered domain. The work on sequence databases is included in the extension of SQL:1999, which supports the notion of window queries over static data streams. Jagadish et al. [11] provided a data model for chronicles (i.e., sequences) of data items and discuss the complexity of executing a view described by the relational algebra operators. The focus of both these efforts was on *stored* time-ordered data rather than on the pipelined processing of live data streams.

Early work on extending database systems to process continuous queries was presented in Tapestry [16], which investigated the incremental evaluation of queries over append-only databases. Tapestry introduces a criterion to judge the correct execution of continuous queries over non-streaming data. The correctness measure in this chapter is an extension of their notion to support sliding window queries.

Window join queries share some similarity with temporal queries. However, efficient algorithms for temporal-joins [13,19] depend on the underlying access structure, which is not available for online stream data sources. Also, the temporal join algorithms do not consider answering continuous queries over infinite data streams.

5.8 Summary

This chapter presented major components of Nile, a prototype data stream management system. These components targeted unique and fundamental data management technologies for streaming applications that range from the flexible representations of streams and sliding window queries (SWQs) to the progressive evaluation of SWQs. Moreover, Nile features two implementations of window operators and their progressive evaluation plans. Both evaluation plans are guided by a simple but concrete correctness measure. One of Nile's main goals is to promote research in data stream management. Towards this goal, Nile was used to research approaches that addressed unique challenges, such as the efficient evaluation of multi-way window joins, the shared execution of continuous queries, the optimized progressive execution of SWQ, and the optimized ordered execution of SWQs. In this chapter we presented two of such research efforts.

References

[1] D.J. Abadi, Y. Ahmad, M. Balazinska et al., The Design of the Borealis Stream Processing Engine. In: CIDR, pp. 277–289, 2005.

[2] W.G. Aref, D. Barbará, S. Johnson, S. Mehrotra, Efficient processing of proximity queries for large databases. In: Proc. of the 11th ICDE, March 1995.

[3] S. Babu, J. Widom, Continuous queries over data streams. In: SIGMOD Record, 30(3), 2001.

[4] P. Bonnet, J.E. Gehrke, P. Seshadri, Towards sensor database systems. In: Proc. of the 2nd Int. Conference on Mobile Data Management, January 2001.

[5] S. Chandrasekaran, O. Cooper, A.D. et al., TelegraphCQ: continuous dataflow processing for an uncertain world. In: CIDR, 2003.

[6] H. Garcia-Molina, J.D. Ullman, J. Widom, Database System Implementation. Prentice Hall, New York, 2000.

[7] T. Ghanem, M.A. Hammad, M. Mokbel, W.G. Aref, A.K. Elmagarmid, Incremental evaluation of sliding-window queries over data streams. IEEE Transactions on Knowledge and Data Engineering, 19(1):57–72, 2007.

[8] G. Graefe, Query evaluation techniques for large databases. ACM Computing Surveys, 25(2):73–170, 1993.

[9] M.A. Hammad, W.G. Aref, A.K. Elmagarmid, Query processing of multi-way stream window join. International Journal on Very Large Data Bases, 2007.

[10] M.A. Hammad, M.J. Franklin, W.G.A. et al., Scheduling for shared window joins over data streams. In: VLDB, pp. 297–308, 2003.

[11] H.V. Jagadish, I.S. Mumick, A. Silberschatz, View maintenance issues for the chronicle data model. In: PODS, May 1995.
[12] J. Kang, J.F. Naughton, S.D. Viglas, Evaluating window joins over unbounded streams. In: ICDE, February 2003.
[13] H. Lu, B.C. Ooi, K.L. Tan, On spatially partitioned temporal join. In: 20th VLDB Conference, September 1994.
[14] P. Seshadri, Predator: a resource for database research. SIGMOD Record, 27(1):16–20, 1998.
[15] P. Seshadri, M. Livny, R. Ramakrishnan, The design and implementation of a sequence database system. In: VLDB, September 1996.
[16] D. Terry, D. Goldberg, D.N. et al., Continuous queries over append-only databases. In: SIGMOD, pp. 321–330, 1992.
[17] P.A. Tucker, D. Maier, T.S. et al., Exploiting punctuation semantics in continuous data streams. TKDE, 15(3):555–568, 2003.
[18] Y. Yao, J. Gehrke, Query processing in sensor networks. In: Proceedings of the CIDR Conference, January 2003.
[19] D. Zhang, V.J. Tsotras, B. Seeger, Efficient temporal join processing using indices. In: ICDE, February 2002.

Chapter 6
Querying of Sensor Data

Niki Trigoni, Alexandre Guitton, and Antonios Skordylis

6.1 Introduction

Advances in micro-electro-mechanical systems (MEMS) allow sensors, actuators, mini-processors and radio devices to be integrated on small and inexpensive devices, hereafter called *sensor nodes*. The deployment of a large number of sensor nodes in an area of interest presents unprecedented opportunities for continuous and untethered sensing in applications ranging from environmental monitoring to military surveillance to disaster relief.

In this chapter, we present a database approach to tasking sensor nodes and collecting data from the sensor network. Similar to the Cougar [35] and TinyDB models [19], we view each sensor node as a mini data repository, and the sensor network as a database distributed across the sensor nodes. As in traditional database systems, users need not be aware of the physical storage organization in order to query data of interest; they should be able to collect sensor data by formulating declarative queries in a language similar to SQL.

Figure 6.1 illustrates a widely used mechanism for tasking sensor networks. A user formulates a declarative query and sends it to the sensor network through a special-purpose node, referred to as the *gateway*. In the *query dissemination phase*, the query is forwarded wirelessly hop-by-hop from the gateway to the relevant sensor nodes. In the *result collection phase*, sensor nodes probe their sensor devices and propagate sensor readings back to the gateway.

N. Trigoni · A. Guitton · A. Skordylis
Computing Laboratory, University of Oxford, Oxford OX1 3QD, UK

N. Trigoni
e-mail: Niki.Trigoni@comlab.ox.ac.uk

A. Guitton
e-mail: Alexandre.Guitton@comlab.ox.ac.uk

A. Skordylis
e-mail: Antonios.Skordylis@comlab.ox.ac.uk

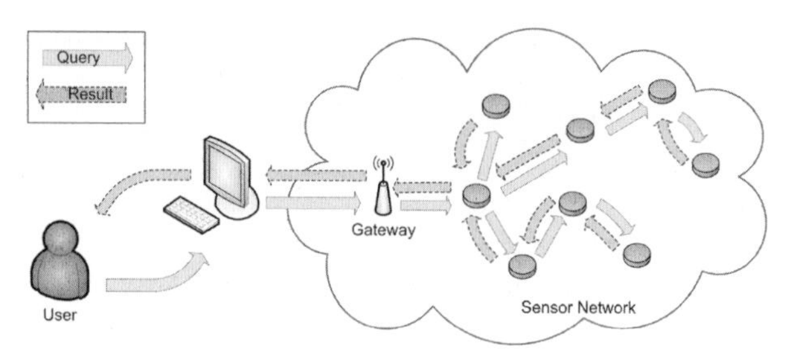

Fig. 6.1 A typical query/response model

The query dissemination phase and the result collection phase must be carefully designed to account for severe communication and energy constraints in sensor networks. Wireless data transmission consumes orders of magnitude more energy than processing on a sensor node [29]. This suggests that battery-powered nodes could significantly increase their lifetime by exploiting their local processing capabilities to minimize the amount of data transmissions. In this chapter, we highlight the importance of *in-network processing*, i.e. evaluating a query inside the network, as opposed to collecting raw data and processing it centrally at the gateway.

The rest of this chapter is organized as follows. Section 6.2 classifies queries into groups based on user-defined quality-of-service requirements. Section 6.3 presents techniques for propagating queries into a sensor network, whereas Sect. 6.4 focuses on two tightly coupled aspects of the data-collection phase, namely routing and in-network processing. Section 6.5 overviews a data-centric approach to storing data within the network, and Sect. 6.6 presents concluding remarks.

6.2 Types of Queries

This section discusses user requirements that arise in various sensor network applications, and provides examples of how they can be expressed using a declarative query language like SQL. We classify queries based on a number of criteria, like frequency, granularity, trigger-mechanism, accuracy and delay of query results.

One-shot vs. long-running queries: We first group queries based on the frequency of requested results: one-shot queries request data from sensor nodes once, as opposed to long-running queries that request results at regular intervals. Consider a sensor network application designed to monitor empty parking spaces in a shopping center. When users arrive, they pose one-shot queries about the state of the parking lot, as follows:

```
SELECT nodeid, loc FROM sensors WHERE space = empty;
```

In contrast, users in environmental applications are typically interested in monitoring an area continuously and receiving data updates at regular intervals. An example of a long-running query, which asks for the temperature value detected by each sensor node every hour, is provided below:

```
SELECT nodeid, temp FROM sensors EVERY 1 hour;
```

`SELECT *` *vs. aggregate queries*: For a large class of applications, like security and surveillance, users are interested in extracting raw sensor data. They use `SELECT *` queries to extract all data from the sensor network or filter it using a predicate, for example:

`SELECT *` FROM sensors WHERE loc in Region;

We distinguish `SELECT *` queries from queries used to obtain aggregate data, e.g. the mean of temperature values in the network. Recent work has focused on common SQL aggregates (like `MIN`, `MAX`, `SUM`, `COUNT` and `AVG`), as well as statistical measures like standard deviation (`STD`) [19]. A user can request aggregate data from different sensors (*spatial-aggregate query*) or from different points in time for a single sensor (*temporal-aggregate query*).

Time-based vs. event-based queries: Consider a long-running `SELECT *` query that asks the temperature of all sensors every hour. This query is time-based since result propagation is triggered by a timer that expires every hour. In the event of a fire, this type of query would prove inadequate since the delay to detect the fire could be prohibitively long. The need for prompt event detection is being met by a class of queries called *event-based*. Event-based queries are typically installed once at the sensor nodes and results are returned when a specified condition is met, for example:

```
ON EVENT temp > 100 SELECT nodeid, temp FROM sensors
                    EVERY 1 sec;
```

Using event-based queries can significantly reduce the amount of energy spent compared to continuously polling the network.

Accurate vs. approximate queries: In large sensor networks, it is not realistic to expect that query results will accurately reflect the current state of the network. Faults naturally occur in a number of places; for instance, during measurement acquisition by sensor devices or in data propagation due to node and link failures. In the light of such uncertainty, users are generally prepared to tolerate various levels of precision. Approximate queries reflect the user's error tolerance by incorporating an error threshold (often followed by a confidence threshold), for example:

```
SELECT nodeid, temp FROM sensors ERROR 2
            CONFIDENCE 95%;
```

The query defines that for each sensor the difference between the reported and real temperature values must be upper bounded by the error threshold 2 with probability 0.95. Tolerating a certain amount of error can significantly reduce the energy spent in result propagation (see [4] and [28]). Accurate queries, are less common, but still useful for safety-critical applications like health monitoring. They express the need for extracting precise sensor data, and suggest trading communication savings for reliable and accurate data delivery.

Urgent vs. delay-tolerant queries: Consider two sensor networks, one deployed for agriculture monitoring and the other for military surveillance. In both networks, users send event-based queries to be notified about interesting events (e.g. changes in soil properties and enemy attacks, respectively). In the agriculture scenario, users might not be pressed for receiving query results immediately after soil changes are observed, whereas in the military scenario the query response time is critical, and data returned late is considered to be stale.

6.3 Query Dissemination

Once the user formulates her query, the query is wrapped into a query message, which is then disseminated wirelessly from the gateway into the sensor network. We describe two distinct approaches to query dissemination. The first approach, referred to as *query broadcasting*, aims to deliver each query to all sensor nodes in the network. The second approach suggests forwarding a query only to those sensor nodes that are likely to contribute to the query result. The latter approach is referred to as *selective query broadcasting*.

6.3.1 Query Broadcasting

This section provides a brief overview of *one-to-all* communication protocols commonly used to disseminate query messages from the gateway to the entire network. An interested reader can refer to [13] or [14] for surveys.

Simple flooding: Early work on query processing for sensor networks focused on long-running queries, for which query messages are disseminated only once, whereas results flow from the sensor nodes to the gateway multiple times at regular intervals [35]. In this case, the cost of query dissemination is negligible compared to the cost of result propagation, and a simple flooding protocol suffices to broadcast query messages to the network.

Simple flooding works as follows. The gateway initially broadcasts the query to all its neighbors. Upon receiving a query message, a sensor node checks whether it is a new query or a duplicate. If the message concerns a new query, the latter is forwarded up to the query processing layer and it is then rebroadcast. The node may introduce a short random delay before rebroadcasting the query in order to avoid collisions due to concurrent transmissions by nearby nodes. The left part of Fig. 6.2 shows an example of simple flooding initiated at the gateway (black node). Each arrow represents the reception of a query message by the target sensor node. Solid arrows indicate reception of queries for the first time, while dashed arrows indicate reception of duplicate queries.

Tree flooding: Simple flooding introduces significant communication overhead, since a large percentage of message receptions involve duplicate queries which are immediately discarded. For example, in the left network of Fig. 6.2 only six message receptions involve new queries (solid arrows) and the remaining 18 receptions

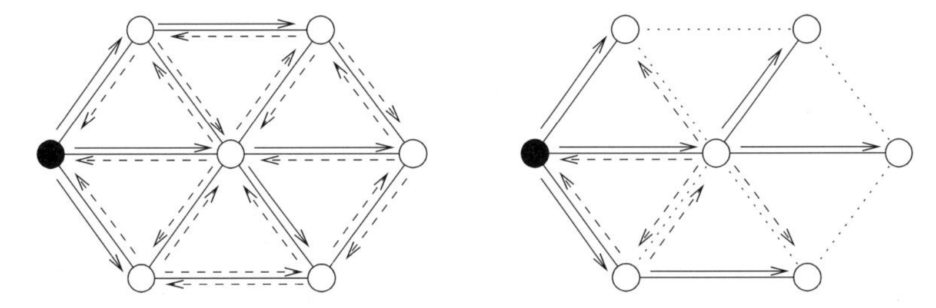

Fig. 6.2 A query is broadcast from the gateway (*black node*) to all sensor nodes in the network using simple flooding (*on the left*) or tree flooding (*on the right*). Each arrow represents the reception of a query message by the target sensor node. Solid arrows indicate reception of queries for the first time, while dashed arrows indicate reception of duplicate queries

involve duplicate ones (dashed arrows). This overhead becomes significant in applications with many one-shot queries for which query propagation is as frequent as result propagation.

Tree flooding can significantly reduce the communication cost by first building a communication tree that spans the entire network, and then propagating queries along the paths of the tree. In scenarios with infrequent link and node failures, the tree is built once, and it can subsequently be used to disseminate multiple queries. Tree construction is performed by flooding tree construction messages from the gateway to all sensor nodes in the network. Upon receiving a tree construction message for the first time, a node sets the sending node as its *parent* and rebroadcasts the message. The node discards any subsequent tree construction messages.

Once a tree is constructed, each sensor node knows its parent on the tree. Queries are disseminated from the root to the leaves as follows: Upon receiving a query message from its parent, a node records it locally and rebroadcasts it. Query messages received from nodes other than the parent are immediately discarded. The right part of Fig. 6.2 shows an example of tree flooding in a small network. The tree is denoted by solid lines. Dotted lines represent edges that do not belong to the tree, and therefore queries disseminated along these edges are discarded. Sensor nodes without children do not need to rebroadcast queries any further. Notice that tree flooding involves only six messages that get discarded at the receiving node, instead of 18 such messages in the case of simple flooding.

Efficient broadcasting: Several energy-efficient broadcasting protocols have been proposed in the literature, which typically aim to reduce the number of nodes that need to retransmit a message [33,27]. Commonly used techniques range from probabilistic broadcasting [37] to geometry-aware [7] and neighborhood-based broadcasting [20]. An example of the latter technique, in which each node knows its two-hops neighborhood, is provided in Fig. 6.3. In both networks, we assume that node s gets to transmit a query message before node t. In the left network, node t decides to rebroadcast the query, in order to reach those neighbors that are not also neighbors of s. In the right network, node t suppresses the query transmission, since all of its neighbors have already received the query message from s.

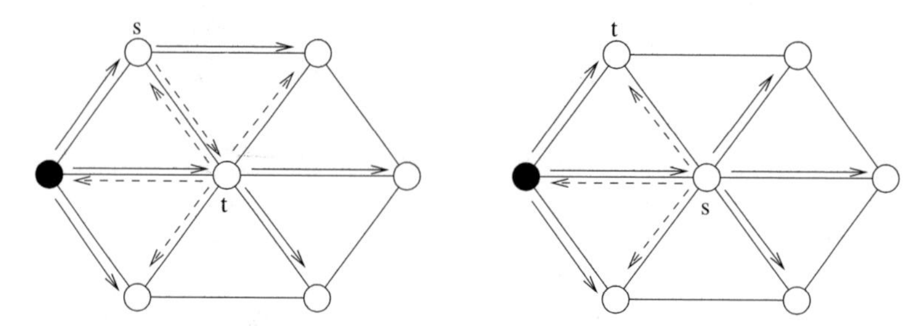

Fig. 6.3 Based on the knowledge of the neighbors of s and of its own neighbors, t decides to retransmit (*on the left*) or not to retransmit (*on the right*)

6.3.2 Selective Query Broadcasting

Selective query broadcasting denotes the forwarding of a query to a specified set of sensor nodes, instead of flooding it to the entire network. For instance, a query about temperature does not need to be forwarded to nodes that are not equipped with temperature sensors. Similarly, a query concerning a specific region does not need to be propagated to sensor nodes that are outside the region, although some of them might be used to forward the query.

Semantic routing trees: In [17], Madden et al. proposed annotating communication trees with semantic information in order to reduce the cost of query propagation. The idea is for a sensor node to avoid disseminating a query to its subtree if it knows that none of its descendant nodes can participate in the query result. A simple and efficient way to implement semantic routing trees is to use a bottom-up approach. Once the tree is built, each leaf sensor node sends to its parent a list of the attributes that it can measure (and potentially value ranges for each attribute). When a sensor node receives attribute lists from all its children, it stores them in memory and sends a merged list to its own parent. Once all sensor nodes in the tree have built their attribute lists, they can easily determine if they have at least one descendant node which can contribute data to a certain query. The approach described in [2] is similar in the sense that each node also maintains a range table for the attributes of its children.

Geographic routing: Geographic routing decreases the cost of query propagation when location information is available at the sensor nodes and queries involve spatial boundaries (e.g. queries that ask for sensor data in a specific region) [16,36,21]. We summarize how to broadcast a query message to all nodes in a target region using GEAR, a geographical and energy-aware algorithm proposed by Yu et al. [36]. We first use greedy forwarding to forward packets to nodes that are always progressively closer to the centroid of the target region, whilst trying to balance energy consumption at the intermediate nodes. Once the query message is delivered to the centroid of the target region, we then use restricted flooding to broadcast the query to all remaining nodes in that region.

6.4 Result Collection

After queries are disseminated into the network, sensor nodes start generating query results. Results are not always returned as soon as queries are disseminated, e.g. in the case of event-based queries. The joint process of processing queries and propagating query results is performed in a distributed manner, and consists of the following steps: Each sensor node receives results from its local sensing devices and from other sensor nodes, which we refer to as *children*. It processes these results locally (Sect. 6.4.1) and forwards them to one or more sensor nodes on the way to the gateway (Sects. 6.4.2 and 6.4.3). The steps of processing and propagating query results are illustrated in Fig. 6.4. By jointly considering the two steps, we can exploit significant opportunities for cross-layer optimization.

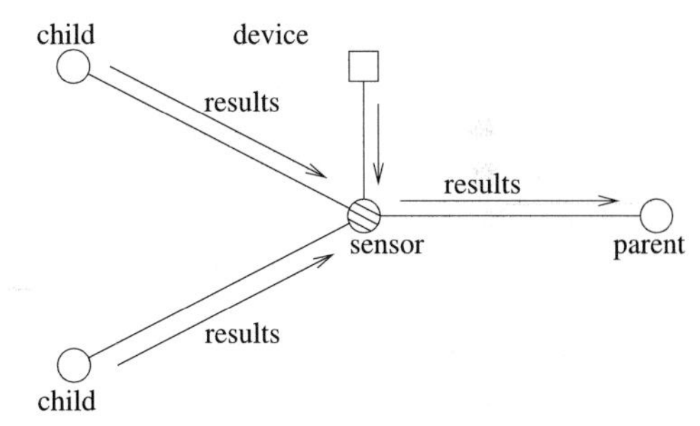

Fig. 6.4 Distributed processing and propagation of the results

6.4.1 In-Network Processing

Inexpensive sensor nodes are equipped with micro-processors with limited processing capabilities. In the light of this constraint, one solution would be for sensor nodes to simply forward their own data and their children's data toward the gateway. However, recent studies on small devices have shown that the energy spent in communicating is orders of magnitude higher than the energy spent in computing [29]. This means that if sensor nodes could locally process query results to reduce the amount of propagated messages, they would be able to make significant energy savings. In what follows, we discuss examples of reducing communication by pushing part of the query processing into the network.

6.4.1.1 Aggregation, Approximation and Compression

One of the most extensively studied forms of in-network processing is in-network aggregation [18,9]. Let us assume that a query requests the aggregate value (e.g. SUM) of an attribute in the entire network. During query dissemination, a communication tree is constructed that connects all sensor nodes to the gateway. Instead of forwarding all attribute values to the gateway to process the query in a centralized manner, it is possible to partially aggregate query results within the network. In-network aggregation involves very simple computations at each node: each node receives partial aggregates from its children, combines them (sums them up) with the local attribute value and forwards the resulting aggregate to the parent node for further processing. Aggregation operators such as `MAX`, `MIN`, `COUNT` and `AVG` can also be processed in a similar way, whereas others like `MEDIAN` are not amenable to distributed processing. An example of a more complex operator, which can be evaluated in a distributed manner, is the union of orthogonal polygons in the context of topology mapping [12].

Another widely used approach to reducing communication cost by processing data within the network is in-network compression. For delay-tolerant queries, a sensor node can defer the propagation of query results to the gateway until it has accumulated enough results to be able to compress them efficiently [11]. For approximate queries, there are further opportunities for communication savings by means of in-network lossy compression. Recent work on in-network processing of approximate queries [4,34] has focused on both reducing communication costs and estimating query results under heavy packet losses.

6.4.1.2 Model-Driven Data Acquisition

Significant communication savings can be achieved by exploiting knowledge of the application domain and the underlying data distributions. In the following, we provide examples of exploiting temporal and spatial correlations in the observed data. The interested reader should refer to [24,3,5,10,11] for further details.

Let us consider a long-running query that requests the temperature measured by a sensor s every 10 minutes. If the temperature measured by s has not changed in the last 10 minutes, s does not need to propagate the new reading. From the absence of readings, the gateway can infer that the temperature remains stable, provided that sensor s occasionally emits heartbeat beacons to denote that it is alive.

In applications where temporal correlations are consistent, queries can be further optimized by reducing the sampling rate of sensor devices (e.g. taking temperature measurements every 20 minutes). In the case of event-based queries, sensor nodes should adapt their sampling rates to the likelihood of an event happening, and to the lifetime of the event. For example, in a wildlife monitoring setting, we need a higher sampling rate to detect cheetahs than turtles, since the latter remain within sensing range for a longer period. An extensive discussion of data acquisition, including sampling rates, can be found in [19].

Spatial correlations can also offer significant communication and sensing savings. Consider the previous example of temperature monitoring, and assume that another sensor t is deployed next to s. Since the two nodes are spatially close, they are likely to generate similar temperature readings. If s and t notify the gateway that there is a strong correlation in their data, then only one of them needs to send temperature reports to the gateway. This node could be the one closer to the gateway, or the one with the longer life expectancy, or both nodes, if they decide to alternate roles to balance their load.

6.4.1.3 Placing Operators in the Network

A query can be represented as a tree of operators that aggregate, correlate, join or filter data streams. Our previous discussion on in-network aggregation assumed operators like SUM and COUNT, which can be evaluated in a distributed manner. We expand our discussion to operators that are not amenable to distributed computation, and highlight the importance of carefully selecting where to place them in the network. Consider, for example, a query that asks for the MEDIAN of humidity values generated by the three nodes at the bottom of Fig. 6.5. Depending on where we place the MEDIAN operator, we will cause different amounts of message traffic. For example, the plan depicted on the left places the MEDIAN operator at the gateway and incurs eight messages, whereas the plans on the middle and right incur six and four messages, respectively. Intuitively, by pushing operators closer to the sensor nodes that contribute to the query results, we can achieve significant communication savings. A more detailed study of adaptive and decentralized operator placement for in-network query processing can be found in [1].

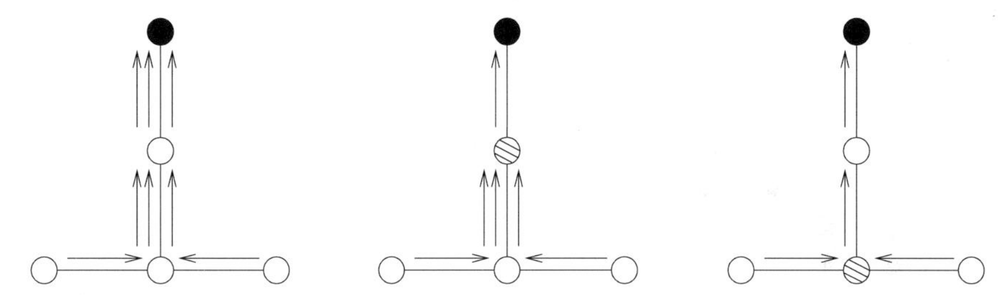

Fig. 6.5 Depending on the location of the MEDIAN operator (indicated by a striped sensor), a different number of messages is generated: eight messages (*on the left*), six messages (*in the middle*) or four messages (*on the right*)

6.4.2 Interplay between Routing and Query Processing

A plethora of recent work has focused on the tight coupling between query processing and routing in sensor networks. As an example, consider the evaluation of aggregate queries over a communication tree. Earlier work [35,19,31] suggested building the tree first, and then partially aggregating results along the paths of the tree. Intanagonwiwat et al. [15] took a similar approach, but they proposed a reinforcement-based algorithm to set up tree routes. Initially, query results start flowing towards the

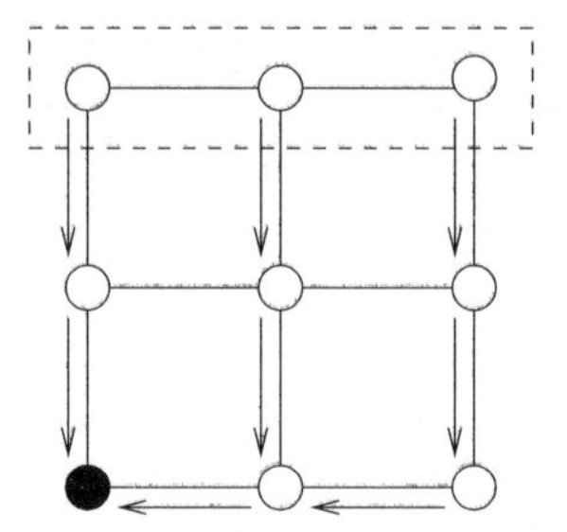
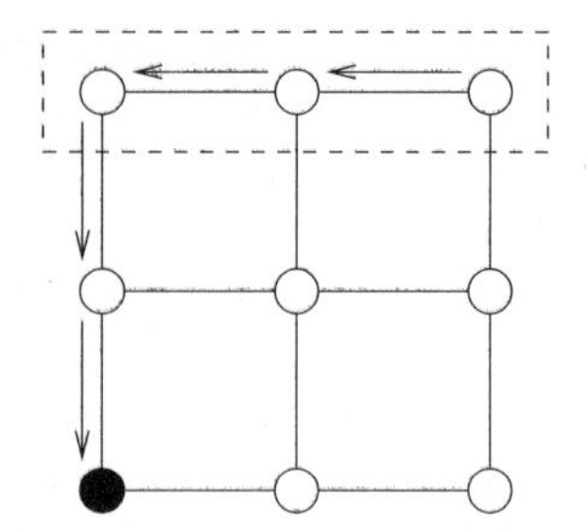

Fig. 6.6 Consider an aggregate query that asks for the sum of all readings in a region (denoted by *the dotted rectangular area*). To collect query results, the left communication tree incurs a cost of eight messages, whereas the right tree aggregates results early and incurs a cost of 4 messages

query initiation point along multiple paths with very small frequency. Depending on their quality, some of the paths are suppressed whereas others are positively reinforced by increasing the frequency of requested results along these paths.

More recent work investigated how the shape of the communication tree may affect the cost of result propagation [30,23]. For example, Trigoni et al. [30] considered the problem of multi-query optimization lifting the assumption of an existing communication tree. The proposed optimization algorithms identify common query sub-aggregates and propose common routing structures to share the sub-aggregates at an early stage. The benefits of carefully selecting a communication tree to forward query results are obvious even for a single query. Figure 6.6 illustrates two ways of processing an aggregate query that asks for the sum of all readings in a region (denoted by the dotted rectangular area). Notice that a total number of eight messages are sent along the left communication tree, whereas only four messages are forwarded along the right tree.

Sharaf et al. [23] proposed a group-aware network configuration method to reduce the cost of GROUP-BY queries. They influenced the construction of the communication tree by *clustering* along the same path sensors that belong to the same group. These later works show that interweaving routing and processing decisions is a fertile area of research in sensor networks, which presents many opportunities for energy-efficient result propagation.

6.4.3 Interplay between Medium Access Control, Routing and Query Processing

Further cross-layer optimizations can be achieved if we try to tune the MAC layer, with the routing and query processing layers. Due to the nature of wireless communication, when two nearby sensors try to transmit at the same time their messages may collide. They will then have to wait for a short period before they try to retransmit. Retries due to collisions lead to higher energy consumption at the nodes, and longer latency in delivering query results. Tens of MAC protocols have recently been proposed in the literature [6]; we focus our attention to TDMA-like protocols, i.e. protocols that schedule nodes to communicate at specified time intervals in order to minimize interference and avoid idle listening. In what follows we provide two examples that show how node schedules can be carefully designed to optimize query execution.

Madden et al. proposed a tree-based transmission schedule for processing aggregate queries in sensor networks [18]. Every node in the tree is allowed to transmit partial aggregates during a specific interval of the query duration, which depends on the node's distance (number of hops) from the gateway. Nodes deep in the tree transmit sooner, ensuring that intermediate nodes higher up in the tree have received results from all descendant nodes before they are ready to partially aggregate them and forward them to their parents. By coordinating node transmissions along the tree, nodes avoid idle listening, and reduce the delay of result propagation. They avoid collisions between nodes at different levels of the tree, but not between sibling nodes at the same level.

Another example of interaction between node scheduling and routing is provided in [25]. The proposed whirlpool data delivery technique divides the network into sectors, and interrogates sensors on a rotating basis to avoid interference among sectors. Whirlpool is combined with an algebraic optimization framework to achieve intra-sector concurrency. The proposed Data Transmission Algebra (DTA) takes as input transmissions of query results and generates a safe schedule of transmissions, which avoids collisions within a sector.

6.5 Data-Centric Storage

In the previous two sections, we discussed a pull-based approach to tasking a sensor network, in which queries are disseminated to pertinent nodes, and results are propagated from these nodes back to the gateway. In applications where queries are known in advance, and arise regularly with high probability, it is preferred to use a push-based approach, in which sensor nodes proactively forward data to the gateway without waiting to be explicitly probed by queries. In this section, we briefly discuss a hybrid push-pull mechanism, in which sensor nodes proactively send data midway into the network, store them in carefully selected storage nodes, from where data is then pulled on demand by propagating queries [32].

Shenker et al. have recently proposed a data-centric approach to storing data within the network [26,22]. The main idea is to store data (raw readings or more composite events) of the same type at the same network location. A hash function is first used to map each data type (e.g. grey elephant sightings) to a key that represents a network location. When data of this type are generated at a source node, it is forwarded to and stored at the node closest to this key location using a geographic algorithm, like GPRS. When a user poses a query that asks for a particular type of data, the query is greedily forwarded towards the key location, and the storage node closest to that location responds with the corresponding data.

In applications where users would like to mine sensor data using drill-down queries, data can be summarized in various resolutions and stored in an hierarchical manner [8]. At higher levels (near the gateway) summaries are highly compressed but correspond to larger volumes of spatio-temporal data, whereas at lower levels, local data is stored at higher resolution. Querying is performed using hierarchical techniques borrowed from data mining: each query is injected at the highest level, where only coarse summaries exist. Queries that ask for low-precision data are processed locally, whereas those asking for high-precision data are forwarded to storage nodes at lower levels. In order to cope with limited storage, an aging algorithm can be used that decides how long each node must maintain summaries at various resolutions before discarding them.

6.6 Concluding Remarks

This chapter provided an overview of the various aspects of query processing for sensor networks. Our premise is that a sensor network can be viewed as a distributed database system and probed using a declarative query language. This allows users to express functional requirements (what data they need) and non-functional requirements (accuracy and delay of results) without worrying about where the data are stored and how they can be accessed.

The severe power and communication constraints inherent to sensor networks have led to the emergence of energy-efficient broadcast protocols for disseminating queries and collecting query results. Since communication is more energy-consuming than computation, most techniques aim to reduce message traffic by abandoning the centralized approach and pushing query operators into the network.

Query evaluation in sensor networks is performed in a distributed manner and is tightly coupled with the underlying routing and MAC layers. Hence, a lot of recent research has focused on cross-layer optimization techniques, like taking into account the query workload to build routing structures and node schedules. Further optimizations can be achieved by combining in-network processing with in-network storage, i.e. selecting a few nodes where data is proactively forwarded and stored, and reactively pulled by queries.

References

[1] B.J. Bonfils, P. Bonnet, Adaptive and decentralized operator placement for in-network query processing. In: IPSN'03: Information Processing in Sensor Networks, pp. 47–62, 2003.

[2] S. Chatterjea, S. De Luigi, P. Havinga, An adaptive directed query dissemination scheme for wireless sensor networks. In: ICPPW'06: International Conference Workshops on Parallel Processing, pp. 181–188, 2006.

[3] D. Chu, A. Deshpande, J. Hellerstein, W. Hong, Approximate data collection in sensor networks using probabilistic models. In: ICDE'06: International Conference on Data Engineering, 2006.
[4] J. Considine, F. Li, G. Kollios, J. Byers, Approximate aggregation techniques for sensor databases. In: ICDE'06: International Conference on Data Engineering, 2004.
[5] A. Deligiannakis, Y. Kotidis, N. Roussopoulos, Compressing historical information in sensor networks. In: COMAD'04: International Conference on Management of Data, pp. 527–538, 2004.
[6] I. Demirkol, C. Ersoy, F. Alagoz, MAC protocols for wireless sensor networks: a survey. IEEE Communications Magazine, 2005.
[7] A. Durresi, V.K. Paruchuri, S.S. Iyengar, R. Kannan, Optimized broadcast protocol for sensor networks. IEEE Transactions on Computers, 54(8):1013–1024, 2005.
[8] D. Ganesan, B. Greenstein, E.D. Perelyubskiy, J. Heidemann, An evaluation of multi-resolution storage for sensor networks. In: SENSYS'03: Embedded Networked Sensor Systems, pp. 89–102, 2003.
[9] J. Gehrke, S. Madden, Query processing in sensor networks. IEEE Pervasive Computing, 46–55, 2004.
[10] C. Guestrin, P. Bodi, R. Thibau, M. Paski, S. Madden, Distributed regression: An efficient framework for modeling sensor network data. In: IPSN'04: Information Processing in Sensor Networks, pp. 1–10, April 2004.
[11] A. Guitton, A. Skordylis, N. Trigoni, Utilizing correlations to compress time-series in traffic monitoring sensor networks. In: WCNC'07: Wireless Communications and Networking Conference, 2007.
[12] J.M. Hellerstein, W. Hong, S. Madden, K. Stanek, Beyond average: toward sophisticated sensing with queries. In: IPSN'03: Information Processing in Sensor Networks. Lecture Notes in Computer Science, vol. 2634, pp. 63–79. Springer, Berlin, 2003.
[13] F. Ingelrest, D. Simplot-Ryl, I. Stojmenovic, Routing and broadcasting in hybrid ad hoc and sensor networks. In: Theoretical and Algorithmic Aspects of Sensor, Ad Hoc Wireless and Peer-to-Peer Networks, pp. 415–426, 2006.
[14] F. Ingelrest, D. Simplot-Ryl, I. Stojmenovic, Energy-efficient broadcasting in wireless mobile ad hoc networks. In: Resource Management in Wireless Networking, pp. 543–582, 2005.
[15] C. Intanagonwiwat, R. Govindan, D. Estrin, Directed diffusion: a scalable and robust communication paradigm for sensor networks. In: MOBICOM'00: Mobile Computing and Networking, pp. 56–67, 2000.
[16] Y.B. Ko, N.H. Vaidya, Location-aided routing (LAR) in mobile ad hoc networks. Wireless Network, 6(4):307–321, 2000.
[17] S.R. Madden, M.J. Franklin, J.M. Hellerstein, W. Hong, The design of an acquisitional query processor for sensor networks. In: COMAD'03: International Conference on Management of Data, pp. 491–502, 2003.
[18] S.R. Madden, M.J. Franklin, J.M. Hellerstein, W. Hong, TAG: a tiny aggregation service for ad-hoc sensor networks. In: OSDI'02: Operating Systems Design and Implementation, pp. 131–146, December 2002.
[19] S.R. Madden, M.J. Franklin, J.M. Hellerstein, W. Hong, TinyDB: an acquisitional query processing system for sensor networks. ACM Transactions on Database Systems, 30(1):122–173, 2005.
[20] W. Peng, X.C. Lu, On the reduction of broadcast redundancy in mobile ad hoc networks. In: MOBIHOC'00: Mobile Ad Hoc Networking and Computing, pp. 129–130, 2000.
[21] A. Rao, C. Papadimitriou, S. Shenker, I. Stoica, Geographic routing with location information. In: MOBICOM'03: Mobile Computing and Networking, pp. 96–108, 2003.
[22] S. Ratnasamy, B. Karp, L. Yin, F. Yu, D. Estrin, R. Govindan, S. Shenker, GHT: A geographic hash table for data-centring storage in sensornets. In: WSNA'02: Wireless Sensor Networks and Applications, September 2002.
[23] M. Sharaf, J. Beaver, A. Labrinidis, P. Chrysanthis, Balancing energy efficiency and quality of aggregate data in sensor networks. The VLDB Journal, 2004.

[24] M. Sharaf, J. Beaver, A. Labrinidis, P. Chrysanthis, TiNA: a scheme for temporal coherency-aware in-network aggregation. In: MOBIDE'03: Data Engineering for Wireless and Mobile Access, pp. 69–76, 2003.
[25] D. Sharma, V.I. Zadoronzhny, P.K. Chrysanthis, Timely data delivery in sensor networks using Whirlpool. In: DMSN'05: Data Management for Sensor Networks, August 2005.
[26] S. Shenker, S. Ratnasamy, B. Karp, R. Govindan, D. Estrin, Data-centric storage in sensornets. In: HOTNETS'02: Hot Topics in Networks, October 2002.
[27] J.P. Sheu, C.S. Hsu, Y.J. Chang, Efficient broadcasting protocols for regular wireless sensor networks: research articles. Wireless Communications and Mobile Computing, 6(1):35–48, 2006.
[28] A. Skordylis, N. Trigoni, A. Guitton, A study of approximate data management techniques for sensor networks. In: WISES'06: Workshop on Intelligent Solutions in Embedded Systems, 2006.
[29] M. Stemm, R. Katz, Measuring and reducing energy consumption of network interfaces in hand-held devices. IEICE Transactions on Fundamentals of Electronics, Communications, and Computer Science, 8(E 80):1125–1131, 1997. Special Issue on Mobile Computing.
[30] N. Trigoni, A. Guitton, A. Skordylis, Poster abstract: Routing and processing multiple aggregate queries in sensor networks. In: SENSYS'06: Embedded Networked Sensor Systems, November 2006.
[31] N. Trigoni, Y. Yao, A. Demers, J. Gehrke, R. Rajaraman, Multi-query optimization for sensor networks. In: DCOSS'05: Distributed Computing in Sensor Systems, pp. 307–321, 2005.
[32] N. Trigoni, Y. Yao, A. Demers, J. Gehrke, R. Rajaraman, Hybrid push-pull query processing for sensor networks. In: Workshop on Sensor Networks, GI-Conference Informatik, 2004.
[33] B. Williams, T. Camp, Comparison of broadcasting techniques for mobile ad hoc networks. In: MOBIHOC'02: Mobile Ad Hoc Networking and Computing, June 2002.
[34] M. Wu, J. Xu, X. Tang, Processing precision-constrained approximate queries in wireless sensor networks. In: MDM'06: Mobile Data Management, May 2006.
[35] Y. Yao, J. Gehrke, The cougar approach to in-network query processing in sensor networks. ACM SIGMOD Record, 31(3):9–18, 2002.
[36] Y. Yu, R. Govindan, D. Estrin, Geographical and energy aware routing: a recursive data dissemination protocol for wireless sensor networks. Technical Report UCLA/CSD-TR-01-0023, University of Southern California, 2001.
[37] Q. Zhang, D.P. Agrawal, Dynamic probabilistic broadcasting in MANETs. Journal of Parallel and Distributed Computing, 65(2):220–233, 2005.

Chapter 7
Aggregation and Summarization in Sensor Networks

Nisheeth Shrivastava and Chiranjeeb Buragohain

Abstract Sensor networks generate enormous quantities of data which need to be processed in a distributed fashion to extract interesting information. We outline how ideas and algorithms from data stream query processing are revolutionizing data processing in sensor networks. We also discuss how sensor networks pose some particular problems of their own and how these are being overcome.

7.1 Introduction

Sensor networks have introduced a new paradigm in distributed data processing systems. Traditional distributed database systems work with a small number of processing units, each highly capable in terms of computation, communication and storage. In contrast, sensor networks consist of a large number of nodes, each with very limited computational and communication capabilities. Moreover, the query processing in sensor network is continuous. Sensors deployed in a certain region individually take continuous measurements of the desired physical phenomenon (e.g. temperature of a building). A typical use of the network is to continuously monitor some statistics (e.g., average or median temperature) over the distribution of values over the entire deployment. The continuous nature of both query and measurements, combined with strict resource constraints, pose many novel challenges in designing algorithms for sensor networks. Query processing techniques in sensor networks must process large amounts of data, generated at distributed locations, and transmit the results back to the user, while consuming very little in terms of resources.

N. Shrivastava
Bell Labs Research, Alcatel-Lucent India, Bangalore, India
e-mail: nisheeths@alcatel-lucent.com

C. Buragohain
Amazon.com, Seattle, WA 98104, USA
e-mail: chiran@amazon.com

A similar trend in resource constrained query processing has arisen in *data stream* systems. A data stream consists of a sequence of data elements which arrive online. The query processing system then has to process these items as they arrive without storing them. Some notable examples of data stream processing include transactions in financial markets and stock exchanges, monitoring and traffic engineering of IP networks, mining large databases etc. The potential number of items in a stream can be unbounded and is typically much larger than the storage capacity of the stream processing system. Therefore storing the data and making multiple passes over it to process queries is impossible in such systems. Measurements at a single sensor can be thought of as a data stream, where the sensor senses the environment periodically and produces a steady stream of unbounded data. Due to its resource constraints, the sensor node can neither store the data locally, nor transmit it to a remote location. A network of sensors therefore can be modeled as a distributed system of data streams over which we need to carry out various queries.

There has been a tremendous amount of work on query processing techniques in data streams. To get a taste of the challenges presented by the data stream model, let us consider two typical queries on a stream of values: average and median. The average can be easily computed online over the data stream by keeping track of only two numbers: the sum of all the values seen in the stream and the number of values in the stream. On the other hand, computing the exact median of n values in a stream is not possible without using $\Omega(n)$ storage [1]. Answering many other useful queries such as counting distinct items, frequent items, top-k, etc. exactly also requires large amounts of memory in the data stream setting. Therefore computing complex queries like median using only small amount of storage requires radically different approach to algorithm design.

The two key ideas on which most data stream processing techniques depend on are *approximation* and *summaries*. Typically, computing *exact* answers to complex queries requires $\Omega(n)$ storage, but if we are willing to accept approximate answers with reasonable accuracy, we can build efficient summaries that answer the same query requiring much smaller storage. For example, although computing the exact median requires $\Omega(n)$ storage, median with ε error can be computed using a summary of size roughly $O(\frac{1}{\varepsilon})$ [13,26]. Therefore in most data stream algorithms, we sacrifice accuracy for the sake of better memory efficiency. Typical data stream algorithms maintain a data structure known as a *summary*, which summarizes the data that has been seen in the stream so far. When a new item in the stream appears, it is used to update the summary and is then discarded. Queries on the stream are answered by querying the summary itself. Although answers from the summary may not be exact, they come with strict theoretical error guarantees. For example, the value reported as median can be guaranteed to have a rank within $\pm 1\%$ of the true median. A data stream algorithm is considered efficient if the size of the summary is bounded by $O(\text{polylog}(n))$ and similarly the update time for any item is also bounded by $O(\text{polylog}(n))$.

This paradigm of summaries and approximation has been very successfully extended to sensor network systems as well. The query processing algorithm first builds a summary of measurements at a single sensor, which is then transmitted over the network to other sensors. These sensors aggregate their own summary into the received summaries and forwards it until it arrives at the sensor node which initiated the query. In keeping with the limited storage and communication and computational resources, a sensor network algorithm is considered efficient if the size of the summary that is processed and transmitted has size bounded by $O(\text{polylog}(n))$. Since sensor networks can be considered as a set of distributed streams, many data stream algorithms have been adapted for sensor network settings. However, since summaries in sensor networks must support the aggregation operation, not all data stream algorithms can be successfully adapted for sensor networks. In fact, sensor networks represent a more general paradigm than traditional data stream systems—each summary in sensor network can also be used in data stream setting; however, the converse is not true.

7.1.1 Aggregation in Sensor Networks

The canonical aggregation framework for sensor networks is tree based. This framework was introduced by Madden et al. [19,20] in the context of TAG: Tiny AGgregation service and demonstrated its utility as a part of a sensor database system known as TinyDB. The TAG framework views the sensor network as a distributed database. The users pose their queries to a special node in the network known as the *base station*, using a SQL-like language. Using the underlying routing protocol of the sensor network, the query is routed to all the nodes in the network. The nodes route their summary back towards the base station along a routing tree rooted at the base station as shown in Fig. 7.1. This routing tree is the backbone around which the aggregation algorithms are organized. As the data flows up the network, the intermediate nodes aggregate these summaries. Therefore this approach is also known as *in-network aggregation*. In this chapter we shall focus on three key challenges in designing efficient aggregation algorithms.

- *Summary for Complex Queries*: How can we compute a summary to answer useful queries like average, sum, median, number of distinct items etc. in an efficient fashion?

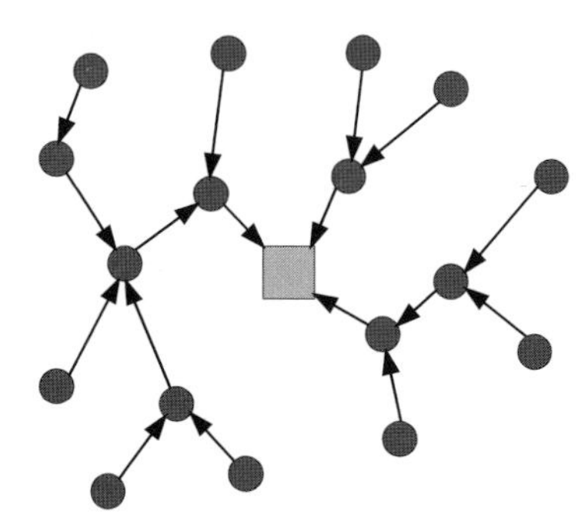

Fig. 7.1 Routing and aggregation tree in a sensor network. The square node at the center is the base station

- *Update of Aggregates*: Since typical sensor network queries are continuous queries, how can we continuously update the query answers efficiently?
- *Fault Tolerance*: Sensor network communication links are notoriously unreliable. Therefore significant challenges appear in designing aggregation algorithms which are resilient to packet loss in the network.

Simple queries like average and min/max can be answered in a sensor network without recourse to complex algorithms. For more complex queries, quantile summaries have proven to be highly versatile. In addition to quantiles, they can answer other queries such as histograms, frequent items and rank queries. Another query of great importance for sensor networks is the distinct item query; especially because the distinct item query algorithm forms the basis on which fault tolerant aggregation algorithms are built. Therefore we shall mostly focus our attention on quantile queries, distinct item queries and how the distinct item query can help answer queries in unreliable sensor networks.

7.2 Preliminaries and Related Work

Let us first start with the model of network that we shall be using. We assume that our sensor network consists of n nodes numbered $1, 2, 3, \ldots, n$. Without loss of generality we assume that each sensor i senses a value x_i which ranges over integers from $[1, \sigma]$. We wish to compute statistics like sum, average, count and median over the set of values $\{x_1, x_2, \ldots, x_n\}$. Therefore we shall use n to simultaneously denote the number of sensors in the network as well as the size of the data set over which we carry out the query. We shall assume that the sensors can form a routing tree through some mechanism like breadth-first search. For simplicity, we shall start by assuming that the wireless links in the routing tree are reliable. We shall later relax the assumption of reliability, and see how that leads to change in aggregation algorithms.

Since all aggregation algorithms in sensor networks can be cast within the summary paradigm, we shall spend a little time in describing how a summary works. A *summary* is a compact data structure that captures the distribution of data in the network such that aggregate statistics about the data set can be computed with ease. Instead of trying to define a summary exactly, we shall give an example which will illustrate the key properties of summaries. Given a summary s, it supports three operations: generation (G), aggregation (A) and evaluation (E). We shall illustrate these operations through an algorithm which can evaluate the count aggregate. The summary s consists of a single integer: $s = \langle c \rangle$. Every sensor i *generates* a summary $s_i \equiv G(i) = \langle 1 \rangle$. The *aggregation* operation is defined by $A(\langle c_1 \rangle, \langle c_2 \rangle) = \langle c_1 + c_2 \rangle$. The *evaluation* operation is defined by $E(\langle c \rangle) = c$.

Then formally we can describe the count evaluation algorithm running on a routing tree as follows: every sensor node generates (G) its own summaries and then waits for summaries to arrive from its children. When all the summaries from the children have arrived, it aggregates (A) those summaries together and forwards the new aggregated summary to its parent. The base station eventually receives all the summaries from its children, aggregates them and carries out the evaluation (E) operation on the final summary to evaluate the count.

The summary structures for evaluating simple aggregates like count, sum, average, min or max are of constant size irrespective of the number of data points n (assuming the count and sum over n values can be represented by a constant size bit vector, e.g., integer). Moreover the answers produced by these summaries are exact. In general, summaries produce answers which are accurate within a fraction ε while consuming storage of the order $O(\text{poly}(\varepsilon^{-1}, \log n, \log \sigma))$.

7.2.1 Related Work

Aggregation techniques in sensor networks have led to a rich set of literature over past few years. TinyDB [20] and Cougar [33] provide database-style query interface to sensor networks and give algorithms for computing various basic summaries such as min, max and average. Zhao et al. [34] also gave summary techniques for min, max and average. However, their techniques are not applicable to complex queries like median, since these queries are not *decomposable*. In [19], Madden et al. claimed that computing holistic aggregates such as median is very hard, and requires sending all the values back to the base station. Most relevant works to the complex queries discussed in this article are distributed summaries in sensor networks for quantile queries. In [26], Shrivastava et al. presented a distributed summary called q-digest, that answers median and other quantile queries with guaranteed error ε. The size of q-digest is bounded by $O(\frac{1}{\varepsilon}\log \sigma)$, which is much smaller than $O(n)$ for TAG [19]. Concurrently, Greenwald and Khanna [14] discussed another summary structure for quantile queries that requires $O(\frac{1}{\varepsilon}\log^2 n)$ memory. Hellerstein et al. [15] proposed approximate techniques to compute contours and wavelets histogram summaries. Even though these summaries are compact and perform quiet well in practice, they do not provide any strict error guarantees. In other recent works, Gandhi et al. [12] and Solis et al. [30] gave space-efficient structures to compute the spatial isocontours (lines that make up the contour map of sensor data), which take into account the spatial correlation of values to reduce the size of summary.

As noted before, there are distinct similarities between answering queries over streaming data and over sensor network values. There is a vast literature in summarizing data streams, and we refer the reader to [3,24] for comprehensive surveys. In a seminal paper [1], Alon, Matias, and Szegedy established lower bounds for a variety of queries over data streams; specifically, they showed that computing quantiles in a single pass over a data stream requires $O(n)$ memory. This can be easily extended to show a lower bound for sensor network model. In [13], the authors gave a space-efficient approximate summary to answer quantile queries. This summary can answer all quantile queries with error εn, and has size at most $O(\frac{1}{\varepsilon}\log n)$. Other works [22] have discussed space-efficient algorithms for finding frequent and top-k items in a data stream. Hershberger et al. [16] presented a data structure to compute median and frequent items over multi-dimensional data in a single pass. Cormode and Muthukrishnan [5] presented a summary which can answer quantile queries under insertion and deletion of values using memory $O(\varepsilon^{-1} \log^2 n \log(\log n/\delta))$ with probability $1 - \delta$.

For continuous queries under updates, Cormode et al. [6] proposed a scheme to answer continuous quantile queries in a distributed data stream context. To minimize the number of updates, they proposed using a prediction model to keep track of errors being accumulated as values change. Babcock and Olston [2] presented techniques to monitor top-k values in a distributed data stream setting. In [27], the authors provided techniques to continuously maintain the quantile summary over sensor values with updates with bounded error. Silberstein et al. [29,28] proposed several energy efficient techniques to monitor top-k and other extreme values under updates. Their scheme uses spatial correlation among update events of neighbors to suppress redundant messages in the network. Following a different approach, Deshpande et al. [7,8] used spatial and temporal correlations in sensor measurements to reduce the data transmitted to base station. They further built models over the measured data and used those models to build cost effective sensing *plans*; i.e., acquire measurements only at selected sensors, hence reducing the amount of data sensed.

Message losses due to unreliable links is one of the biggest sources of error in sensor network data collection. In [4,25], the authors suggested using multi-path routing topology for aggregation which is more resilient to link losses than the tradition tree-based routing. Considine et al. [4] discussed how to compute min, max and average in this multi-path routing model, which is robust to link losses. Nath et al. [25] formalized this notion of robustness into a framework which allows one to handle packet losses more gracefully. The summary computation over multi-path routing is more heavyweight and requires much more communication than a tree topology. Motivated by this, Manjhi et al. [21] presented a hybrid routing approach, that starts off with tree routing in part of network farther away from base station, and switches to multi-path as the messages gets close to base station.

7.2.2 Roadmap

We begin our discussion in Sect. 7.3 with the problem of finding complex aggregates like median over the sensor network. We present two approaches to this problem, the q-digest (Sect. 7.3.1) and the Greenwald–Khanna algorithm (Sect. 7.3.2). In the context of q-digest we also discuss the problem of continuously updating aggregates over the network using small amount of computation. In Sect. 7.3.3, we discuss the problem of estimating the number of distinct values over the network which has important implications for designing other sensor network algorithms. Then in Sect. 7.4 we discuss the question of how to accurately compute aggregates like average and count in the face of packet loss within the network. Finally in Sect. 7.5, we discuss some of the unsolved problems in the field and possible future directions.

7.3 Complex Queries in Sensor Networks

Given the set S of values sensed by the sensor $\{x_1, x_2, \ldots, x_n\}$, the approximate distribution of these values represents a fundamental query over the sensor network system. Although traditional equi-width histograms are highly useful in describing a data distribution, in many realistic cases histograms do not capture important aspects of the data. An equi-depth histogram [23] which can answer percentile (or quantile) queries can be a much richer description of the data distribution. Therefore, in this section, we focus on the following basic quantile query: given a number q between 0 and 1, what value has a rank of qn in sorted sequence of values in S? Using the quantile summary, one can answer a variety of useful queries such as rank and range queries, frequent items, histogram etc. Since this is such a fundamental problem, we shall present two very different deterministic algorithms to solve this problem: the quantile digest algorithm (Sect. 7.3.1) and the GK algorithm (Sect. 7.3.2). Finally we look at the distinct item query in Sect. 7.3.3 which counts the number of distinct values in the set $\{x_1, x_2, \ldots, x_n\}$.

7.3.1 *The Quantile Digest Algorithm*

In this section we present the data structure called q-digest (quantile digest), which captures the distribution of sensor data approximately. The core idea behind q-digest is to divide the range of values into variable-sized *buckets* according to the data distribution such that each bucket contains an almost equal number of values. The buckets in q-digest are chosen from a binary partition (represented as binary tree T) of the input value space $1, \ldots, \sigma$ as shown in Fig. 7.2. Without loss of generality, we shall assume that σ is a power of 2. For example, the root bucket in T corresponds to the bucket $[1, \sigma]$. The left and right children of the root correspond to the buckets $[1, \sigma/2-1]$ and $[\sigma/2, \sigma]$, respectively. In general, each bucket $v \in T$ corresponds to the value range $[v.\min, v.\max]$ and maintains a variable count(v)

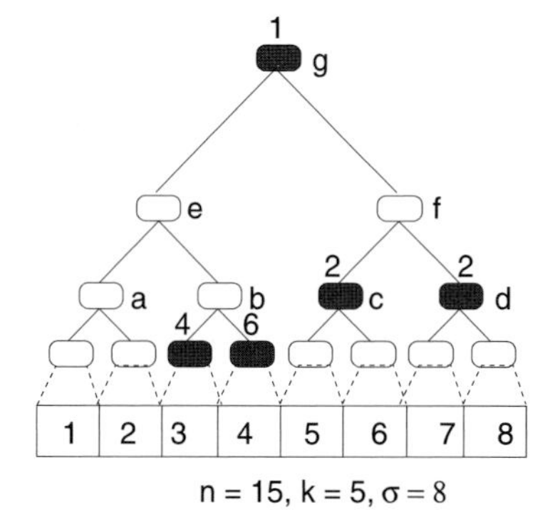

Fig. 7.2 q-digest: Complete binary tree T built over the entire range $[1, \ldots, \sigma]$ of data values. The bottom most level (leaf level) represents single values. The dark buckets are included in the q-digest Q, and number next to them represent their counts

which counts number of values tracked in this bucket. The count(v) counts a subset of values in the range: each value x is counted in exactly one bucket, but this bucket could be any one of the $\log n$ buckets whose range contain x. This property is very unlike traditional histograms where a single bucket counts *all* the values in its range.

The q-digest is a subset of these possible buckets with their associated counts. Since we are constrained in memory, the q-digest keeps track of the data distribution using only $O(k)$ buckets, where we call k the *compression parameter*. A bucket v is in q-digest if and only if it satisfies the following *digest property*:

$$\text{count}(v) \leq \lfloor n/k \rfloor, \tag{7.1}$$

$$\text{count}(v) + \text{count}(v_p) + \text{count}(v_s) > \lfloor n/k \rfloor, \tag{7.2}$$

where v_p is the parent and v_s is the sibling of v.

The properties work towards making buckets of roughly equal counts (intuitively, like equi-depth histogram). The first constraint asserts that no bucket should have a very high count (more than n/k). In contrast, the second says that there should not be adjacent buckets (bucket, its sibling and parent) with very small count (less than n/k). If there are such buckets, we merge them into the parent bucket to reduce memory. Formally, a q-digest satisfying these properties have at most $3k$ buckets, which makes it highly space efficient.

7.3.1.1 Building a q-Digest

Let us first look at construction of q-digest over a set of values present at a single location, we will discuss the distributed construction a little later. Given a set S of n values, we first make a trivial q-digest Q containing only leaf level buckets of T, and set the count of bucket v as the frequency of value $v.\min$ in S. We then apply the following *compress* operation on the q-digest: we go through the buckets bottom up and check if any bucket v violates the digest property 7.2. If we find such a bucket, we remove v and its sibling from Q and add their counts to the parent of v. The process ends when we reach the root bucket. An example of q-digest is shown in Fig. 7.2

In a sensor network the q-digest is constructed in a distributed fashion, using in-network aggregation. The process starts at the sensor having no children in the routing tree, where a single value can be considered a trivial q-digest with one leaf-level bucket. Each sensor node collects q-digests from its children sensor in the routing tree and then aggregates them into a single q-digest. To aggregate (A) q-digests, we take the union of the q-digests and add the counts of buckets with the same range ([min, max]). Then, since the size of q-digest may have increased, we apply *compress* operation to the resulting q-digest. At the base station (which is the root of the routing tree), we get a q-digest containing values of all sensors in the network.

7.3.1.2 Queries and Analysis

Q-digest is a versatile summary that captures the distribution of data, which can answer a variety of queries such as: quantiles, inverse quantiles (rank query), histogram and consensus values. Here we will concentrate only on the quantile query, and refer the reader to [26] for details about other queries. In a quantile query, the aim is the following: given a fraction $q \in (0, 1)$, find the value whose rank in sorted sequence of the n values is qn. The q-digest can answer quantile queries with error at most ε: if the returned value has true rank r, then the error $|r - qn| < \varepsilon n$.

To answer the quantile queries, we do a *post-order* traversal on Q, and sum the counts of all the buckets visited before a bucket v. We report the value $v.\max$ as qth quantile, for which this sum c becomes greater than (or equal to) qn. The post-order traversal ensures that all values included in c are surely less than $v.\max$. But notice that some values counted in ancestors of v (buckets whose range contains the range of v) can also be less than $v.\max$, which won't be included in c. These values are the source of error in the quantile query, since these should be counted in the true rank of v, but are not. Using q-digest properties, and choosing $k = \frac{1}{\varepsilon} \log n$, we can show that this error is bounded by at most εn.

7.3.1.3 Continuous Query and Updates

Thus far we have discussed how to build the quantile summary over measurements taken by the sensors at a given time. Many practical settings, however, warrant a *continuous* monitoring of the phenomenon being measured. Some examples are wildlife habitat monitoring [31], tracking environmental variations [18], structural monitoring [32] etc. Under such settings the user not only requires accuracy of the summary, but is also interested in the adaptability—the summary must represent the *most recent* measurements. This means that the q-digest summary must be periodically updated as the underlying physical phenomenon changes. We assume that each sensor in the network takes periodic measurements at specific (time synchronized) intervals, called *epochs*. The user is interested in finding the summary of measurements taken in the current epoch.

The q-digest data structure does not support *delete* or *replace* operations, hence modifying the q-digest partially as some values change is non-trivial and could require almost as much communication as building it from the scratch. Instead, we look at an alternate way to reduce communication cost, by reducing the number of recomputes on the q-digest. The basic theme is to delay recomputing as much as possible, without violating the error guarantee. Notice that now there can be two sources of error—due to summarization and due to delay in recomputing. If the user specifies a maximum error of ε, the initial q-digest is build with error $\varepsilon/2$ and the rest of $\varepsilon/2$ error is left for the lazy updates. We will discuss three adaptive update schemes for maintaining q-digest, namely Value-Adaptive, Range-Adaptive and Area-Adaptive, as described in the following.

In the Value-Adaptive scheme, we count the number of sensors whose measurements have changed since the last digest computation. If this count is less than $\varepsilon n/2$, we can be sure that the digest at the root has an error of at most εn. If it becomes more than $\varepsilon n/2$, we recompute the q-digest. This simple scheme reduces the communication at every epoch, but still performs lots of unnecessary re-computations, because the number of value updates is not a true indicator of the potential increase in error. For example, if a value change does not change count of any bucket in the q-digest of the base station, the original guarantees are still valid, hence the recomputation was unnecessary.

In the Range-Adaptive scheme, the base station transmits the final q-digest Q ranges back to every sensor. If the new value at a sensor belongs to the same range in Q as the previous value, the value update creates no extra error. However, if the new value belongs to a different range, it could result in an increase in error at the base station. We call such value update a *jump update event*, and count such events in the network. If the number of jump update events is less than $\varepsilon n/2$, the error in q-digest is at most εn. Otherwise, we recompute the digest. Even though this requires twice the amount of communication at every recompute, in practice it reduces the total communication cost drastically.

The third scheme, Area-Adaptive, reduces the communication cost of q-digest recomputation by *localizing* it to the portions of network that have actually changed. Each sensor counts the jump update events in its subtree and while recomputing the q-digest, if that count is less than $\varepsilon n_1/2$, where n_1 is the size of its subtree, then it suppresses the recomputation in its subtree. In this manner, the sensors in the region where phenomenon has not changed significantly will not recompute the q-digest and reduce the communication cost.

7.3.2 The Greenwald–Khanna Quantile Algorithm

We will now discuss another quantile summary called GK summary presented by Greenwald and Khanna in [14]. GK summary, like q-digest, can answer quantile queries with guaranteed error and is of bounded size. The GK summary builds upon a technique presented by the same authors [13] for computing quantile summary in a centralized data stream setting. It is a very good example of how data stream techniques can be helpful in designing efficient data structures for sensor network.

Given a set S of values, the GK summary is a subset $Q = \{q_1, q_2, \ldots, q_t\}$ of S, where $q_1 \leq q_2 \leq \cdots \leq q_t$. Each value has two rank associations, $r\min(q_i)$ and $r\max(q_i)$, which are respectively the minimum and maximum possible rank of q_i in the sorted sequence of S. The first and last element in Q, q_1 and q_t, are the smallest and largest value in the input set S. Since we know the rank exactly for these values, we set $r\min(q_1) = r\max(q_i) = 1$ and $r\min(q_t) = r\max(q_t) = n$. The GK summary has the following properties.

1. For all values $q_i \in Q$, the true rank of q_i is at least $r\min(q_i)$ and at most $r\max(q_i)$.
2. For all i, $r\max(q_{i+1}) - r\min(q_i) < 2\varepsilon n$.

Intuitively, the second property bounds the error in quantile query by making sure that for any given rank r, Q contains at least one value whose true rank is in the range $[r - \varepsilon n, r + \varepsilon n]$. Further, using the first property, we can get that value: find q_i such that $r \in [r\min(q_i), r\max(q_i)]$. Notice that there can be more than one (but at most two) values in Q, whose rank range contains r; both are valid answers to the quantile query. A GK summary of size $t = O(\frac{1}{\varepsilon}\log n)$ can answer quantile query with error at most εn [13].

The quantile summary Q_v for a sensor node v is of the form $Q_v = \{Q_v^1, Q_v^2, \ldots, Q_v^k\}$. Here Q_v^i is a GK summary for a subset of n_i values, where $2^i < n_i \leq 2^{i+1}$. The authors call the ith summary belongs to the ith *class*. Since there are at most $k = \log n$ classes, for any node v, the size of summary Q_v is at most $\log n \cdot O(\frac{1}{\varepsilon}\log n) = O(\frac{1}{\varepsilon}\log^2 n)$. The reason of having multiple classes of summary is to bound the error in aggregating summaries which will be clear a little later.

7.3.2.1 Building a GK Summary

The summary computation in sensor network model occurs in the same bottom-up fashion in the routing tree over sensor nodes. To explain the construction of the GK summary in a sensor network, we simply show how to aggregate (A) GK summaries of two sensors. To aggregate GK summaries Q_u and Q_v, we scan them in increasing order of classes and *aggregate* the summaries in the same class, until no two summaries have the same class. Observe that we can always aggregate two summaries of class i to give a summary of class $i+1$, hence the aggregation always terminates with at most one summary in each class. The aggregation of two GK summaries consists of two operations, merge and compress, defined as follows.

- *Merge*: To merge two GK summaries Q_1 and Q_2 into a single summary Q, we simply take the union of values of the two summaries, sort them and redefine the functions $r\max$ and $r\min$ so that they reflect the true min and max rank of values. Specifically, for value x in the union that originally belonged to the summary Q_1, let a and b be consecutive values in Q_2 such that $a < x < b$. Then, we define $r\min_Q(x) = r\min_{Q_1}(x) + r\min_{Q_2}(a)$, and $r\max_Q(x) = r\max_{Q_1}(x) + r\max_{Q_2}(b) - 1$.
- *Compress*: After the merge operation, the size of GK summary would be $|Q| = |Q_1| + |Q_2|$, which could be undesirably high. To bring the size back to $O(\frac{1}{\varepsilon}\log n)$, we make a compressed GK summary Q' as follows: query Q for ranks $1, n/B, 2n/B, \ldots, n$ and include the values found as the result for each query into Q', along with their $r\min$ and $r\max$ values. It is easy to see that the size of final summary $|Q'| = B + 1$. Also, if the error in Q is ε, the error in Q' is at most $(\varepsilon + 1/(2B))$ [14].

7.3.2.2 Queries and Analysis

To answer the quantile query at the root node, we first merge GK-summaries for all classes into a single GK-summary Q and then use Q to answer the query. Specifically, for quantile query q, we return the value $v \in Q$, such that $r\min(v) < qn < r\max(v)$. We now look at the error memory trade-off of GK summary.

The aggregation operation bounds the size of GK summary, but introduces an extra error of $1/(2B)$ per summary. If a single GK summary is kept at each sensor node, due to repeated aggregation in the network this error could potentially become linear in network size (imagine a routing tree in form of a linear chain of n nodes). This is why we keep $\log n$ classes of GK summaries. To understand the motivation behind keeping these classes, let us see the following claim: for any GK summary Q in class i, the total distinct compress operations performed on values included in the summary in class j is at most $2^{i-j} - 1$, for $j < i$. The reason is that there are only $2^i/2^j = 2^{i-j}$ possible summaries of class j that were aggregated to form Q, which amounts to $2^{i-j} - 1$ total aggregations. Further, the total error due to aggregations in class i is (number of aggregations times $1/(2B)$ times size of summary) at most 2^i. Adding this for all $j < i$, the error due to aggregation is at most $i \cdot 2^i \cdot 1/(2B) \le \frac{\varepsilon}{2}n$, for $B = \frac{1}{\varepsilon}\log n$ and $i \le \log n$. Setting the error in each summary Q^i as $\varepsilon/2$, the total error in quantile query is εn.

The q-digest and GK algorithms solve the same problem, but they are very different in flavor. Although the memory bound of GK $O(\frac{1}{\varepsilon}\log^2 n)$ is not directly comparable to the q-digest bound of $O(\frac{1}{\varepsilon}\log\sigma)$, typical data stream usage assumes that σ and n are polynomially related. For a typical sensor network, assuming $n = 1024$, $\sigma = 2^{16}$, the Greenwald and Khanna memory bound translates to $\frac{1}{\varepsilon} \times 100$, while the q-digest bound translates to $\frac{1}{\varepsilon} \times 16$. Of course, this is a very crude estimate and ignores constant factors hidden in $\mathcal{O}$ notation. On the flip side, the q-digest algorithm assumes that the value are integers, while the GK algorithm can function with arbitrary values. Only actual implementation on real hardware can probably decide the suitability of one algorithm over the other in practice. Incidentally, there exists other quantile algorithms which can be applied to sensor networks and data streams. The Count-Min sketch by Cormode and Muthukrishnan [5] is a probabilistic summary which can answer quantile queries within an accuracy of ε using memory $O(\varepsilon^{-1}\log^2 n \log(\log n/\delta))$ with probability $1 - \delta$. But clearly this algorithm is not competitive with the deterministic algorithms described earlier and will not be discussed.

7.3.3 The Flajolet-Martin Distinct Item Summary

Given the set $\{x_1, x_2, \ldots, x_n\}$, the distinct item query returns the number of distinct values in the set. This is one of the fundamental queries supported by the TinyDB system [19]—moreover it forms the basis on which aggregation algorithms for unreliable sensor networks are built. The obvious algorithm of keeping a list of all distinct items does not scale very well because if the number of distinct items is $O(n)$, then the list will occupy $O(n)$ memory. The first space-efficient summary for this problem was designed by Flajolet and Martin [11]. Suppose we have a set of elements with duplicates such that the total number of unique elements is N. Then the Flajolet-Martin (FM) summary allows us to estimate the value of N within a relative error of ε, using space $O(\varepsilon^{-2} \log N)$. In fact, this bound is known to be close to the best achievable [17], i.e. any algorithm which can estimate the value of n within a relative error ε must use space $\Omega(\varepsilon^{-2})$. Although there are several variants and optimizations of the FM summary [10,9], for the sake of brevity we discuss the simplest variant here.

Consider a set S of elements with duplicates and a "good" hash function h which maps any given element in S to a random number between 0 and 1. Assuming that the hashed values are independent and uniformly distributed, the probability that any given element x hashes to a value less than 2^k is given by $1/2^k$. Let us define the function $\rho(x)$ to be the rank of the first 1 bit in the binary expansion of $h(x)$, e.g. if $h(x) = 0.0010011$, then $\rho(x) = 3$. In other words, the probability that $\rho(x) = k$ is $1/2^k$. Suppose we hash all the elements in S and find that $\rho(x) = 1, 2, 3, \ldots, k$ were all observed. In that case it is fair to say that the total number of distinct elements in the set S is close to 2^k. To keep track of $\rho(x)$, we create a bitmap M with $\log N$ bits where N is the expected number of distinct elements. Initially all the bits in M are set to 0. If $\rho(x) = p$, then we set the pth bit from left of M to 1. Suppose the number of consecutive 1 bits in M starting from the left after hashing the full set S is k. Then an estimate of the number of distinct elements in S is given by 2^k. Suppose we had two sets S_1 and S_2 and constructed the corresponding FM summaries $M(S_1)$ and $M(S_2)$. Then it is not hard to see that the FM summaries for the union of two sets is just the bitwise OR of the two summaries, i.e.

$$M(S_1 \cup S_2) = M(S_1) \vee M(S_2). \tag{7.3}$$

The estimate for distinct elements that the FM summary provides is prone to large amounts of error. For example, an error of ± 1 in ρ can lead to an incorrect estimation of N by a factor of two. To increase the accuracy of this estimate, we can keep m different copies of M and then average over the different estimates to scale the error down to $(1/\sqrt{m})$. Therefore the final aggregation algorithm is as follows: given an error limit ε, every sensor creates an FM summary with $O(1/\varepsilon^2)$ bitmaps, each bitmap of size $O(\log n)$. The sensor node inserts its value x_i into the summary and forwards it to its parent. When a sensor node receives a set of summaries from its children, it aggregates them together with its own summary by carrying out a bit-wise or of all the corresponding bitmaps. The base station can finally carry out the average and output an estimate for N. The maximum size of any message that a sensor needs to transmit is $O(\varepsilon^{-2} \log n)$ for a given relative error ε. To put these numbers in perspective, let us assume that the network contains 1,000 nodes and we wish to achieve a 10% error. Then the size of the summary is close to $10/0.1^2 = 1{,}000$ bits, or 125 bytes.

Note that unlike the q-digest or the GK algorithm, this distinct item algorithm is a probabilistic algorithm: it does not always return the correct answer, but the probability of failure is low. In addition to answering the distinct values query, the FM summary is crucial in constructing aggregation algorithms which are resistant to packet loss. It is to this topic that we turn to in the next section.

7.4 Aggregation in Lossy Networks

Many sensor network aggregation protocols make the assumption of network reliability. But real-life sensor networks do not operate with reliable transmission protocols. Aggregation trees are extremely vulnerable to packet loss: if the message sent by a sensor node to its parent is lost, then the information contained in the complete subtree is lost. Thus packet loss near the base station can be truly catastrophic to the value of the final aggregate.

There are two ways to address this shortcoming: the first and more traditional method is to implement a reliable transport protocol such as TCP over the network and ensure that packet loss doesn't occur over any link. An alternative is a simpler protocol like directed diffusion which uses multi-path routing to mitigate the effect of loss. In a multi-path routing protocol, packets are not routed along a tree, but instead the source sends the packets to all its neighbors and the neighbors in turn forward it towards the base station. Thus multiple copies of a packet are forwarded towards the base station with the assumption that even in the face of loss, at least one of them will arrive.

Considine et al. [4] and Nath et al. [25] suggested using such a multi-path routing protocol to aggregate data in sensor networks. Although multi-path routing seems to evade the problem of data loss, it comes with its own problems. Consider the problem of estimating the count aggregate in a sensor network as described in Sect. 7.2. When we use a tree-based routing scheme, every node forwards one message to its parent which contains the number of nodes in its subtree. This algorithm is not extensible to multi-path routing directly, because with multi-path routing, the count of a subtree might reach the base station multiple times and this will lead to overcounting.

7.4.1 Summary Diffusion in Sensor Networks

To design a fault-tolerant aggregation algorithm for sensor networks, Nath et al. [25] introduced the idea of *summary diffusion* under multi-path routing schemes. We compute a *breadth-first-search* tree rooted at the base station over the communication graph and assign every node a *depth* which is equal to its hop distance from the base station. The summary generation and aggregation algorithm is very similar to the case of tree routing. Every sensor node generates (G) its own summaries and then

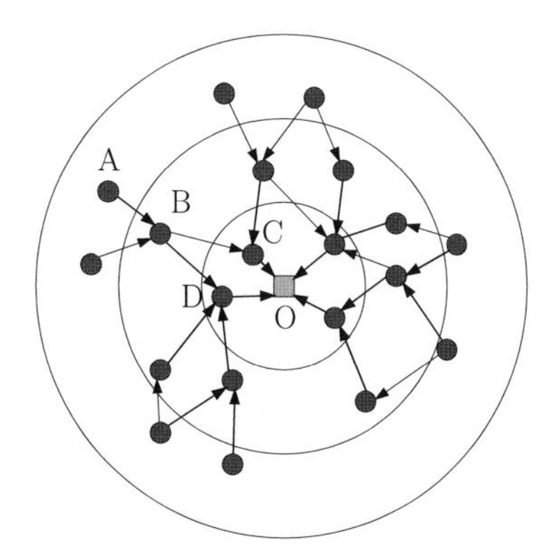

Fig. 7.3 Summary diffusion in sensor networks. The circles define the communication range of sensors. Every node forwards its summary to its neighbor who resides in the inner circle. The summary from node A can reach the base station via two routes: ABCO and ABDO

waits for summaries to arrive from its neighbors who are further from the root than itself. When all the summaries from the neighbors have arrived, it aggregates (A) those summaries together and forwards the new aggregated summary to its neighbors who are closer to the base station than it (see Fig. 7.3). The base station eventually receives all the summaries from its neighbors, aggregates them together and carries out the evaluation (E) operation on the final summary to evaluate the count.

7.4.2 The Duplicate Insensitive Property

The key difference between the tree routing algorithm and the summary diffusion algorithm is the possibility that in summary diffusion, multiple copies of a summary might propagate through the network and hence can be merged together. Therefore if multiple copies of the same summary are merged together, they should not affect the final summary. It can be shown that the summary diffusion algorithm will produce the correct answer as long as the summary is duplicate insensitive and order insensitive. Formally, given two summaries s_1 and s_2, the aggregation operation A must satisfy the following properties.

$$\text{Order Insensitiveness: } A(s_1, s_2) = A(s_2, s_1), \tag{7.4}$$

$$\text{Duplicate Insensitiveness: } A(s_1, s_1) = s_1. \tag{7.5}$$

The traditional summaries for min or max are already duplicate insensitive and hence require no modification for summary diffusion algorithms. But for statistics like count, sum and median, the traditional summaries are no longer sufficient and in the next few sections we look at some of the proposed summaries to compute these statistics.

7.4.3 Duplicate-Insensitive Count and Sum Summary

The problem of counting the number of sensor nodes in the network can be mapped to the distinct counting problem as follows: given that every node has a unique id, how can we count the number of unique ids in the network in the face of duplicates? Considine et al. [4] adapted the FM summary described in Sect. 7.3.3 to solve the counting problem. From the property 7.3 of the FM summary, it immediately follows that the FM summary is order and duplicate insensitive. Therefore the duplicate insensitive count problem is solved by simply creating the FM summary for distinct items, where each sensor inserts its unique id i into the summary. At the base station, the number of distinct items in the FM summary is the number of sensors in the network.

The FM summary can be extended to compute the sum aggregate in the following fashion. Suppose the sensor node i is sensing an integer value x_i. In creating the count summary, the node carries out a single insertion to the summary: its own id. Likewise in the sum summary, the node can carry out x_i distinct inserts to the summary while taking care that these inserts are unique to it. Then the final summary that arrives at the base station will output the sum aggregate. The drawback of this algorithm is that the insertion in any node will take $O(x)$ time. Considine et al. [4] gave a more efficient algorithm based on this idea which takes $O(\log^2 x)$ time. In the interest of brevity we shall not discuss this algorithm in detail and instead refer the reader to the original work.

In addition to sum and count, quantiles are obviously important aggregates that need to be computed in lossy networks. The simplest duplicate-insensitive algorithm to compute the median is random sampling. According to "folk theorems", a random sample of size $O(\varepsilon^{-2} \log \delta^{-1})$ can answer a median query within an error of ε with probability at least $1 - \delta$. Since even computing the count aggregate requires $\Omega(\varepsilon^{-2})$ memory for lossy networks, it seems unlikely that the median can be computed with memory less than $O(\varepsilon^{-2})$. Nevertheless, the problem of deterministically computing median in lossy networks using $O(\varepsilon^{-2})$ memory still remains open.

7.4.4 The Tributary-Delta Approach

From the earlier discussion, it is clear that using multi-path routing increases robustness, but at a large cost in communication overhead because of the requirement of maintaining duplicate sensitive summaries. Also multi-path routing introduces its own approximation errors in statistics like count. Tree routing is more efficient in terms of native accuracy and size of summary, but its limitations are exposed by packet loss.

The Tributary-Delta (TD) approach attempts to combine both tree routing and multi-path routing in the same network to increase reliability and accuracy without too much increase in communication overhead. The key insight of the TD approach is that the effect of packet loss is most severe near the base station because loss of one message can lead to loss of data coming from a large number of nodes. Therefore the TD approach to aggregation uses tree routing far from the base station and switches to multi-path routing close to the base station. The name Tributary-Delta is given in analogy with a river which flows from distant regions to the sea. Away from the sea, the river collects water from many tributaries akin to tree routing, while near the sea the river splits into many streams within a delta akin to multi-path routing. The key question in the TD approach then, is how do individual sensor nodes decide whether to take part in tributary-style tree routing, or delta-style multi-path routing? Manjhi et al. [21] designed algorithms which adaptively shrink or grow the delta region to maintain a desired accuracy level. In the interest of space, we shall not discuss the algorithms and instead refer the reader to the original paper.

7.5 Conclusion and Future Directions

Aggregation in sensor network is a very fast-evolving field. We discussed one of the key shortcomings of the aggregation algorithms, that they ignore the fact that sensor data has strong spatio-temporal correlation. Two sensors geographically near each other produce similar values. Similarly, a single sensor which is continually monitoring a physical variable typically produces a stream of values which are correlated in time. Aggregation algorithms which exploit such correlations [7,12] can significantly cut down on the amount of processing and communication. In this chapter we talked mostly about aggregating numeric data; sensors can generate many other types of data, though, such as event data. Acquiring and processing such event data can itself be a challenge. As sensor technology improves, due to Moore's law we can expect to see the amount of storage and computation available on a single node increase, but the cost of communication will remain almost the same. So we can expect to see new aggregation algorithms which can exploit large amounts of storage and computation available on nodes.

References

[1] N. Alon, Y. Matias, M. Szegedy, The space complexity of approximating the frequency moments. In: STOC '96: Proceedings of the Twenty-Eighth Annual ACM Symposium on Theory of Computing, pp. 20–29, New York, NY, USA. ACM Press, 1996.
[2] B. Babcock, C. Olston, Distributed top-k monitoring. In: SIGMOD '03: Proceedings of the 2003 ACM SIGMOD International Conference on Management of Data, pp. 28–39, New York, NY, USA. ACM Press, 2003.
[3] B. Babcock, S. Babu, M. Datar, R. Motwani, J. Widom, Models and issues in data stream systems. In: PODS '02: Proceedings of the Twenty-First ACM SIGMOD-SIGACT-SIGART Symposium on Principles of Database Systems, pp. 1–16, New York, NY, USA. ACM Press, 2002.

[4] J. Considine, F. Li, G. Kollios, J. Byers, Approximate aggregation techniques for sensor databases. In: Proc. of the 20th Intl. Conf. on Data Engineering (ICDE), 2004.
[5] G. Cormode, S. Muthukrishnan, An improved data stream summary: the count-min sketch and its applications. In: Proc. of LATIN 2004, 2004.
[6] G. Cormode, M. Garofalakis, S. Muthukrishnan, R. Rastogi, Holistic aggregates in a networked world: distributed tracking of approximate quantiles. In: Proc. of SIGMOD'05, 2005.
[7] A. Deshpande, C. Guestrin, W. Hong, S. Madden, Exploiting correlated attributes in acquisitional query processing. In: Proc. of International Conference on Data Engineering (ICDE 2005), 2005.
[8] A. Deshpande, C. Guestrin, S. Madden, J. Hellerstein, W. Hong, Model-driven data acquisition in sensor networks. In: Proc. of the 30th International Conference on Very Large Data Bases (VLDB 2004), 2004.
[9] M. Durand, P. Flajolet, Loglog counting of large cardinalities. In: European Symposium on Algorithms (ESA03), 2003.
[10] P. Flajolet, Counting by coin tossings. In: Lecture Notes in Computer Science, vol. 3321, pp. 1–12, 2004.
[11] P. Flajolet, G.N. Martin, Probabilistic counting algorithms for data base applications. Journal of Computer and System Sciences, pp. 182–209, 1985.
[12] S. Gandhi, J. Hershberger, S. Suri, Approximate isocontours and spatial summaries in sensor networks. In: International Conference on Information Processing in Sensor Networks (IPSN'07), 2007.
[13] J.M. Greenwald, S. Khanna, Space-efficient online computation of quantile summaries. In: Proc. the 20th ACM SIGMOD Intl. Conf. on Management of Data (SIGMOD), 2001.
[14] J.M. Greenwald, S. Khanna, Power-conserving computation of order-statistics over sensor networks. In: Proc. of 23rd ACM Symposium on Principles of Database Systems (PODS), 2004.
[15] J.M. Hellerstein, W. Hong, S. Madden, K. Stanek, Beyond average: toward sophisticated sensing with queries. In: F. Zhao, L. Guibas (Eds.), Information Processing in Sensor Networks. Springer, 2003.
[16] J. Hershberger, N. Shrivastava, S. Suri, C.D. Toth, Adaptive spatial partitioning for multidimensional data streams. In: Proc. of the 15th Annual International Symposium on Algorithms and Computation (ISAAC), 2004.
[17] P. Indyk, D. Woodruff, Tight lower bounds for the distinct elements problem. In: Proc. of the 44th IEEE Symposium on Foundations of Computer Science (FOCS), 2004.
[18] James reserve microclimate and video remote sensing, http://www.cens.ucla.edu.
[19] S. Madden, M.J. Franklin, J. Hellerstein, W. Hong, Tag: a tiny aggregation service for ad-hoc sensor networks. In: Proc. of OSDI '02, 2002.
[20] S. Madden, S. Szewczyk, M.J. Franklin, D. Culler, Supporting aggregate queries over ad-hoc sensor networks. In: Workshop on Mobile Computing and Systems Application, 2002.
[21] A. Manjhi, S. Nath, P.B. Gibbons, Tributaries and deltas: efficient and robust aggregation in sensor network streams. In: Proc. of SIGMOD'05, 2005.
[22] G. Manku, R. Motwani, Approximate frequency counts over data streams. In: Proc. 28th Conf. on Very Large Data Bases (VLDB), 2002.
[23] M. Muralikrishna, D.J. DeWitt, Equi-depth histograms for estimating selectivity factors for multi-dimensional queries. In: SIGMOD Conference, pp. 28–36, 1988.
[24] S. Muthukrishnan, Data streams: algorithms and applications, 2003.
[25] S. Nath, P.B. Gibbons, S. Seshan, Z.R. Anderson, Synopsis diffusion for robust aggregation in sensor networks. In: Proc. of SenSys'04, 2004.
[26] N. Shrivastava, C. Buragohain, D. Agrawal, S. Suri, Medians and beyond: new aggregation techniques for sensor networks. In: Proc. of SenSys'04, 2004.
[27] N. Shrivastava, C. Buragohain, D. Agrawal, S. Suri, Continuous quantile queries in sensor networks, 2007.
[28] A. Silberstein, R. Braynard, J. Yang, Constraint chaining: on energy-efficient continuous monitoring in sensor networks. In: SIGMOD '06: Proceedings of the 2006 ACM SIGMOD

International Conference on Management of Data, pp. 157–168, New York, NY, USA. ACM Press, 2006.
[29] A. Silberstein, R. Braynard, C. Ellis, K. Munagala, J. Yang, A sampling-based approach to optimizing top-k queries in sensor networks. In: ICDE '06: Proceedings of the 22nd International Conference on Data Engineering (ICDE'06), p. 68, Washington, DC, USA. IEEE Computer Society, 2006.
[30] I. Solis, K. Obraczka, Efficient continuous mapping in sensor networks using isolines. In: MOBIQUITOUS '05: Proceedings of the Second Annual International Conference on Mobile and Ubiquitous Systems: Networking and Services, pp. 325–332, Washington, DC, USA. IEEE Computer Society, 2005.
[31] R. Szewczyk, A. Mainwaring, J. Polastre, D. Culler, An analysis of a large scale habitat monitoring application. In: Proc. of SenSys '04, 2004.
[32] N. Xu, S. Rangwala, K. Chintalapudi, D. Ganesan, A. Broad, R. Govindan, D. Estrin, A wireless sensor network for structural monitoring. In: Proc. of SenSys '04, 2004.
[33] Y. Yao, J. Gehrke, The cougar approach to in-network query processing. ACM SIGMOD Record, p. 9, 2002.
[34] J. Zhao, R. Govindan, D. Estrin, Computing aggregates for monitoring wireless sensor networks. In: The First IEEE Intl. Workshop on Sensor Network Protocols and Applications (SNPA), 2003.

Chapter 8
Sensory Data Monitoring

Rachel Cardell-Oliver

Abstract The goal of sensory data monitoring is to maximise the quality of data gathered by a sensor network. The principal problems for this task are, specifiying which data is most relevant to user's goals, minimising the cost of gathering that data, and clearing the gathered data. This chapter outlines the state-of-the-art in addressing each of these challenges.

A sensor network is a flexible and powerful instrument for monitoring complex real-world systems at a temporal and spatial granularity not previously possible. Sensory data monitoring generates data sets from which new temporal and spatial patterns can be discovered, leading to better understanding of the monitored environments. The goal of sensory data monitoring is to maximize the quality of gathered data within the constraints imposed by the sensor network instrument. This chapter describes state-of-the-art techniques for sensory data monitoring.

Traditionally, scientists have performed sensory data monitoring using individual sensor data loggers, each periodically collecting and saving data at a specific location. At the end of an experiment loggers are collected and their data downloaded and analysed in the laboratory. A crucial difference between sensor networks and traditional data loggers, is that a sensor network is *programmable*: each node in the network runs a computer program that controls the sensing, filtering, storage and transmission of data. Because a sensor network is programmed, it is able to adapt its behavior to current conditions in its environment. For example, it can filter, discard, store, aggregate, compare with historical data, and label the data it gathers. Furthermore, sensor network nodes are equipped with short-range radio, enabling them to communicate with each other and to transfer their data in real time to web gateways. This makes immediate analysis of the data possible and opens up opportunities for user interaction with the sensor network. A sensor network can vary its

R. Cardell-Oliver
School of Computer Science & Software Engineering and Cooperative Research Centre for Plant-based Management of Dryland Salinity, The University of Western Australia, Perth, Australia
e-mail: rachel@csse.uwa.edu.au

sensing frequency over time and space, as well as varying its data processing and reporting strategies as circumstances dictate.

Another difference between a sensor network and a traditional logging system is the relatively low cost of sensor nodes. A sensor network can have many more monitoring points than a traditional system, giving it the ability to provide a finer-grained picture of a landscape. As a consequence, a sensor network can also generate orders of magnitude more data points than would be gathered in a traditional logging field trial. Sensor networks are often heterogeneous, with different nodes performing different data-collection activities depending on their capabilities, resources and their current role within the network.

A sensor network is built from a collection of nodes, where each node has very limited processing and energy resources. The power and flexibility of a sensor network comes from the combined behavior of many individual monitoring nodes. However, the resources of individual nodes must be carefully managed in order to optimize the outcomes of sensory data monitoring. Nodes run on battery power, possibly replenished by solar panels. In most cases sensing does not require large amounts of energy. However, transmitting data, or even listening for neighborhood data, is a significant consumer of battery power. Thus, the use of the radio for data transmission must be strictly rationed if a network is to be able to run unattended for long periods. Furthermore, nodes are subject to failure and error and so sensor network designs must use diversity and redundancy to overcome the inherent faultiness of their components.

Any scheme for sensory data monitoring must satisfy the *goals* of the sensor network's users. What properties or patterns of the data are the users interested in? What trade-offs are they prepared to make for gathering that data given the constraints of a sensor network instrument? Having identified end-user goals, we are then able to design a sensor network that delivers data of the highest quality and significance for the user. Although this sounds obvious, many sensor network projects have failed because they did not start by designing an application that maximizes the value of sensor network data for its users.

The problem of sensory data monitoring can be characterized as follows: to design a network of sensor nodes to measure selected environmental variables in order to construct a temporal and spatial map of the behavior of the environment according to the goals of the sensor network's users. Both time and space are continuous over the real numbers $\mathcal{R}$. The real-world system under observation can be viewed as a three-dimensional spatial landscape ($\mathcal{L} = \mathcal{R}^3$), observed over time ($\mathcal{T} = \mathcal{R}$), for a vector of environmental variables ($\mathcal{V}$). Values for environmental variables are taken from the space $\mathcal{W}$ of possible sensor values. The environment $\mathcal{E}$ being monitored by a sensor network can thus be modeled as a partial function from landscape, time and environmental variables into values. Equivalently, $\mathcal{E}$ is a subset of the cross-product of these spaces.

$$\mathcal{E} \subseteq (\mathcal{L} \times \mathcal{T} \times \mathcal{V} \times \mathcal{W}).$$

Ideally, sensor networks collect a set of discrete observation points $\mathcal{O}$ from their environments, giving:

$$\mathcal{O} \subseteq \mathcal{E}.$$

It is this observation set $\mathcal{O}$ that is used by scientists to explore behaviors, answer questions and test theories about the behavior of the environment. The data points gathered do not necessarily allow full reconstruction of the underlying space, but rather the data must be *fit for purpose* of the analysis required by the scientists.

In reality, the data gathered by a sensor network is noisy and so we do not immediately have access to true observations $\mathcal{O}$. The task of sensory data monitoring is complicated by problems of erroneous readings and missing data from sensor streams [22,15]. We use $\mathcal{A}$ to denote the actual, gathered data set containing dirty data. A significant task for sensor network design is to recover clean data $\mathcal{O}$ from the actual data $\mathcal{A}$. The underlying environment model $\mathcal{E}$ is assumed to be the ground truth, that is the "true" readings for this system.

The problem of data stream cleaning has three parts. First, to identify the subset of erroneous readings within $\mathcal{A}$. Second, to identify the true values corresponding to the erroneous readings. Third, to extend the set of readings by adding true values for readings that are missing from $\mathcal{A}$.

We can now define the three principal problems for sensory data monitoring:

- *Specifying* the set $\mathcal{O} \subseteq \mathcal{E}$ of data that is relevant to user goals;
- *Minimizing* the cost of gathering the actual data set $\mathcal{A}$ whilst *maximizing* its value to users; and
- *Cleaning* the actual data set $\mathcal{A}$ to recover a cleaned set of observations $\mathcal{O} = \text{clean}(\mathcal{A})$.

The sensory data monitoring problem for sensor networks is difficult because many traditional fault tolerance solutions, essentially over-sampling, are not suitable for sensor networks. In particular, sensor networks have limited energy and bandwidth to deliver data and limited memory to store it, thus indiscriminate over-sampling to overcome noise and data loss is not feasible. Furthermore, there are many points of failure in the chain of processing, storage and communication used by a sensor network to deliver data from a sensor to a lab-based data repository, and so tracing errors and recovering from failures can be difficult.

In this chapter we describe techniques for addressing the sensory data monitoring problem, focusing on three areas: specifying *which* data is of interest to users, identifying *how* nodes will select (only) that data, and identifying and correcting *errors* in gathered data.

8.1 Specifying Sensory Data Monitoring Goals

In this section, we consider methods for expressing *what* data the user is interested in. In general, this area remains a challenging, open research problem, but fortunately there are a number of useful, partial solutions available. Figure 8.1 shows a typical sensor data set with seven soil moisture data streams monitored over two days, including a significant rainfall event on day 126. The full environment is shown in the figure. Soil hydrologists who use this data are primarily interested in behavior near the rainfall event, as well as in relative differences between the data streams from different locations. This data set contains only a small amount of noise.

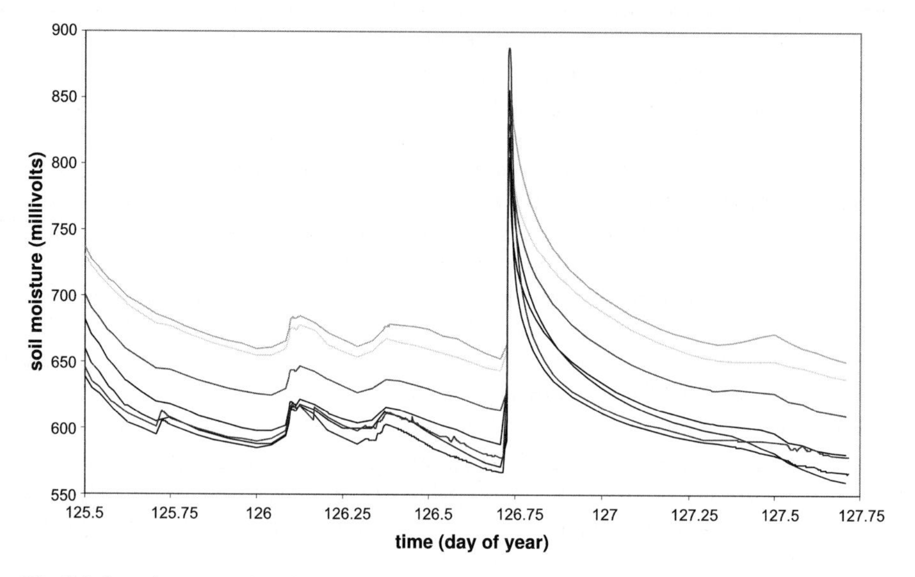

Fig. 8.1 Sample sensory data for soil moisture

8.1.1 Data Management Tools

Early research on designing and building sensor networks focused on issues of transporting data from sensor nodes to permanent storage where it is available for analysis. Analysis software is run on powerful user machines, without the memory, bandwidth, processing and storage limitations of sensor network nodes. The MOTE-VIEW tool integrates data storage, data visualization (by spreadsheet, time series plots or spatial maps), network health management, and calibration [42]. MOTE-VIEW is intended to be a general purpose tool for different applications, and so, for example, provides general solutions for visualizing large data sets.

For end users, however, the focus is usually on the information that can be derived from the gathered data rather than the process of obtaining that data, and so application-specific tools are necessary for effective analysis [35]. For example, to meet end-user requirements we have developed a Soil Moisture Analysis tool that addresses specific issues associated with soil moisture analysis and the derivation of intrinsic soil properties, particularly unsaturated hydraulic conductivity [35].

TASK is a tool that provides end user support for TinyDB applications [7]. Its functions include node health monitoring, software deployment and network reconfiguration, and converting sensor readings to standard engineering units. TASK provides data visualization tools. However, the problems of cleaning data and specifying what data are needed are left to the underlying application software, in this case TinyDB.

Although end-processing of sensor network data is attractive for its flexibility and the power available for processing, there are several disadvantages with this approach, including the high cost of transporting and delivering low-value or erroneous data, and the difficulty of cleaning and selecting relevant data from very large data sets once gathered. For these reasons researchers have investigated ways of moving some of the processing and analysis work earlier in the data gathering chain: at the nodes where the data is sensed. In the remainder of this chapter we describe methods for distinguishing high value data from low value, and for distinguishing correct sensor readings from erroneous ones. However, before we can do that, the user must have a way of expressing what data they are looking for. We review the state of the art in this area using both intentional and extensional programming languages.

8.1.2 Intentional Programming Languages

Intentional programming languages enable the programmer to state *what* is required, rather than how that requirement is to be achieved. The intentional approach is ideal for sensor networks whose users are likely to be experts in the application domain, but not experts in the computing, networking and engineering domains that are necessary for writing effective extensional programs. Intentional programming languages are supported by compilers or interpreters that take user requirements and automatically generate programs that achieve those requirements.

TinyDB and related systems, are based on database technologies [30,20,29,5, 38,21]. They view a sensor network as a database that can be queried. Database tables correspond to the readings of a particular sensor type and table rows correspond to individual readings taken at different times and locations. To interact with a TinyDB sensor network, users write SQL-style queries that are compiled into tasks and injected into the network where the task gathers and returns data to the user. For example, a user can request the average value of all temperature readings above a threshold value from all nodes in a certain area.

The query language of SQL has limited support for both temporal and spatial properties in databases. An alternative and richer approach is provided by the active database model of *situations* [1,27]. Situations are patterns on a set of events that can express properties such as the temporal ordering of events, the absence of an event or the number of times it occurs within an interval of time. SENSID is a middleware implementation of temporal situation detection designed specifically for sensor networks [23]. The model of temporal situations can be further extended to a full space-time logic: a bi-modal requirements logic based on timed propositional temporal logic and a one-hop spatial modal logic [11]. This highly expressive requirements logic can express both spatial (e.g. isolines) as well as temporal (e.g. event order) patterns on events. Developing effective algorithms for compiling space-time logic into tasks to be executed in the network is the subject of ongoing research.

8.1.3 Extensional Programming Languages

Although the intentional approach offers many advantages for sensor network programming, it has proved difficult to design general-purpose intentional languages suitable for a wide range of applications. As a result, almost all existing approaches to programming sensor networks are extensional. That is, the user writes protocol programs stating *how* data should be gathered and transmitted by each node in the network. Here we review some high-level extensional programming methods that have been recently proposed for sensor networks to ease the application designer's task.

Maté is a virtual machine framework for sensor network programming [26]. Maté provides support for users to build application-specific virtual machines, and is also optimized to generate highly efficient run-time code for sensor networks. Complex low-level code for operations or services are represented by a single high-level primitive named by an opcode. Opcodes are a type of application programmer interface to complex low-level functions. Virtual machine code is inserted into event handlers, called contexts, each of which is a sequence of opcodes. For example, using the TinyScript language for Maté, a context for sensor readings repeatedly triggered by a repeating timer could be written:

```
t1 = gettempsensor;
if (relevant(t1)) then store(t1);
if (urgent(t1)) then transmit(t1);
sleep
```

Context scripts such as this one can be inserted into the network at run-time, allowing users to adapt the high-level behavior of the network after deployment. However, the size of applications is limited since the context's opcode programs have to be sufficiently small to transmit over the network in a few packets. Furthermore, in practice it is difficult to keep contexts on all nodes up to date under the unreliable operating conditions of sensor network deployments. Agilla is an extension of the Maté virtual machine to support mobile code fragments that migrate around a sensor network [16,19]. SensorWare has a similar aim, but is designed for higher powered nodes [6].

The goal of *abstract regions* is to simplify sensor network design by providing a set of reusable, general-purpose communication primitives to support local in-network processing [44]. Regions within a network are specified in terms of radio connectivity, geographic location or other properties of nodes. Abstract regions provide programmer interfaces for identifying neighbors, sharing data, performing efficient reductions on a set of data readings as well as providing feedback to users on the resource consumption and level of accuracy of region operations.

An alternative abstraction for sensor network programming is *abstract channels* [37]. In this approach, applications are specified as a set of periodic tasks that can read or write to channels. A channel may be a local sensor or actuator, a local neighborhood of nodes, or a multi-hop channel between, say, a data gathering node and a base station. In each case, the implementation of the channel is performed automatically at run-time, freeing the user from the complex details of this task, and also from the details of code allocation in heterogeneous and dynamically changing networks with mobile nodes. In the latter respect, the approach is similar to role-based programming for sensor networks [17].

EnviroSuite provides an object-oriented model that can encode spatial abstractions on events within a group of nodes [28]. Kairos is a global macro-programming model for sensor networks providing abstractions for node lists, neighbor lists and remote data access for shared variables [18]. In ATaG a sensor network application is driven by its data [2]. ATaG applications are specified as a set of periodic or event driven tasks communicating over abstract channels. DFuse users specify task graphs, where tasks merge multiple data input streams, such as video data, from different locations [24].

8.2 Identifying Significant Data: In-network processing

Intentional and extensional languages both provide a way for users to express their goals for sensor network data collection. We now consider specific mechanisms for the efficient processing of that data within the sensor network.

In-network processing is the general term used for techniques that process data on a node or group of nodes before forwarding it to the user. The goal of in-network processing of data streams is to select and give priority to reporting the most relevant data gathered. The main motivation for in-network processing is that sensor networks do not have sufficient bandwidth, storage or reliability to deliver every data point in time and space. A secondary motivation is to reduce the burden of data overload on the end-user.

Most schemes for compressing data streams introduce some delay in delivering the data, since several data points are collected and processed before being forwarded to other nodes. For many applications this delay is acceptable and well justified by the savings made in the cost of reporting data. For data streams with urgent events that require immediate reaction by the network or an end user, reporting urgent data simply raises an exception to the standard compression mode. For example, while data stream values remain within safe thresholds, then all data is stored, compressed and forwarded with a delay, whilst any data values outside the normal range are reported immediately. For some types of data, such as images or sound, useful compression may be done as soon as the phenomena is sensed, without delaying data reporting. A survey of data-specific compression methods such as image compression, however, is beyond the scope of this chapter. In the following we present both temporal and spatial data compression techniques, and also an approach for utilizing mobile nodes to gather only the most relevant data.

8.2.1 Temporal Compression

A typical series of data readings of natural phenomena such as temperature, rainfall or soil moisture, will contain intervals where there is little change, interspersed with periods of significant change in the value of that variable. In general, scientists will be most interested in the *dynamics* of the monitored variable: how it changes over time. This suggests that a simple way of reducing the cost of reporting sensory data streams is to transmit only the significant changes.

There are several well established techniques for efficiently representing the dynamics of a data stream including lightweight linear compression [36,10], and Haar wavelets [46]. In order to be useful in a sensor network, a compression algorithm must not only reduce significantly the amount of data to be transmitted, but must also be simple to process on each node. We have experimented with several algorithms, and found that linear compression provides the best balance between maximizing compression and minimizing the on-node cost of that compression, and so we focus on that algorithm only in this chapter. Further details can be found in [33].

We now describe a basic algorithm for linear compression that may be varied for different applications [10,33]. Linear compression seeks to represent a time series of readings r_i taken at regular times t_i where $\forall i . t_{i+1} = t_i + \delta$ for a fixed time interval δ. The algorithm selects a subset of these (r_i, t_i) pairs as follows. Any subsequence $(r_j, t_j), (r_{j+1}, t_{j+1}, \ldots, (r_{j+k}, t_{j+k})$ can be approximated by only its endpoints $(r_j, t_j), (r_{j+k}, t_{j+k})$ if and only if all the intermediate points lie within a give error threshold, ϵ, of the line segment between the endpoints.

The algorithm for calculating a linear compression of a data series is a greedy sliding window algorithm. That is, each reading is processed as it is received. Readings are taken at a regular interval δ. For fault tolerance, readings are stored at least every Δ interval, even if there has been no significant change during that interval. Each data series is associated with an error tolerance ϵ: only differences greater than ϵ are considered significant. The error tolerance may be fixed throughout the monitoring process, or may change over time dependent on other environmental variables (e.g. the presence or absence of rain when measuring soil moisture), or the resources available (e.g. battery power on a node or number of nodes monitoring in the same region), or chosen by user preference. By changing the error threshold we can make trade-offs between the accuracy of the selected subset of readings, and the cost of transmitting those readings.

For each monitored time series, the first reading (r_0, t_0) is saved in the node's logging area. As further readings are taken, the ultimate and penultimate sensed values are retained, last $= (r_j, t_j)$ and penult $= (r_{j-1}, t_{j-1})$ as is the value of the last saved reading, saved $= (r_i, t_i)$. The difference between pairs of readings is used to determine whether to save or discard each incoming reading. To account for an accumulation of small errors causing a change exceeding the threshold, we also compare the latest reading with the last stored reading. By default, all readings are logged at least once per latency period.

1. If $|\text{last} - \text{penult}| > \epsilon$, then save last and save penult
2. Else if $|\text{last} - \text{saved}| > \epsilon$, then save last
3. Else if $t_j - t_i > \Delta$, then save last

Figure 8.2 demonstrates the effect of in-network processing using piecewise linear compression. One hundred and twenty soil moisture sensor readings are shown

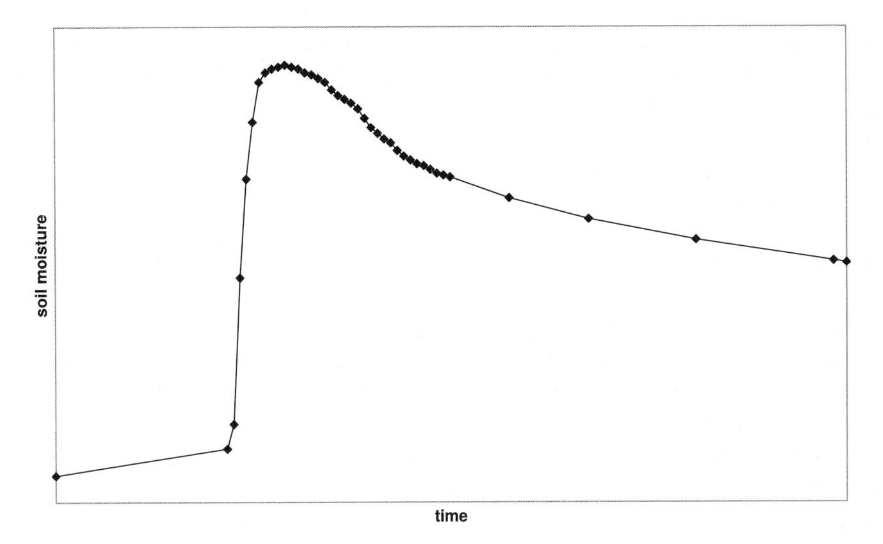

Fig. 8.2 Piecewise linear compression of soil moisture data

over a two-hour period. The underlying series (shown by the solid line) is read at one-minute intervals ($\delta = 60$ seconds) and the error threshold is set conservatively, to half the published error margin (30 mV) of the Decagon soil moisture sensor [13] being used. Logged data are shown by the markers on the line.

The linear compression algorithm has been tested on a variety of data sets from our sensor network field trials [10]. For data sets of 132 to 2208 soil moisture readings, linear compression reduced the sets by at least 76%, with a best case of 95%. For six of the nine data sets tested, linear compression achieved reductions of more than 94% [10]. Moss also reports compressions of 93% to 98% for soil moisture [33].

Soil moisture is a good candidate for linear compression since its data series typically have long periods of little change, interspersed with major changes during and after rainfall. It turns out that linear compression is also highly effective for environmental data series such as temperature, humidity, and wind speed, that undergo more continuous change over time. For example, linear compression was shown to provide reductions of 44% for temperature, 56% for humidity and 64% for wind speed [33]. Even with a 0% error threshold, linear compressions of up to 58% are achieved [33].

8.2.2 Spatial Compression

Over-sampling is often used in sensor networks to provide fault-tolerance to node failures and errors. When there are few failures, then within the sampling landscape the measured data streams will include repetitions of the same or similar values.

A well-known technique for in-network compression of these data is to aggregate similar data values from the same physical area, and to send a single aggregate value rather than all the individual readings. There are many possible aggregation functions for a set of readings including one or more of the mean, the mode, the range, minimum and maximum [29]. Aggregation is the subject of Chap. 7 and so will not be discussed further here.

Just as we can report an efficient representation of a temporal data stream by simply reporting the dynamics (significant changes) within that stream, so we can give an efficient spatial representation by reporting only the *isobars* of a set of spatial readings at any time [44,11]. That is, if the readings of a node and its neighbor differ by more than some threshold, then one of those nodes (say the node with the higher reading) reports the two readings and locations of the nodes. Otherwise, if a node's readings are similar to those of all its neighbors, then that reading is from a plateau and need not be reported.

8.2.3 Mobile Sampling

Most compression schemes assume a sensor network of static sensors that are over-sampling their environment. Spatial and temporal compression are used to filter the gathered data. An interesting alternative approach is to use mobile sensors that are able to focus their sampling in areas where the underlying data streams are most interesting. Batalin et al. [3] introduced *fidelity driven sampling* for monitoring phenomena that require high spatial density sampling to achieve a specified accuracy of environment reconstruction. They proposed sensor networks of both static and mobile nodes. Mobile nodes are tasked to perform high-fidelity sampling in areas of greatest interest. Static nodes perform initial, coarse-grained readings of the environment, and then allocate tasks to perform finer grained sampling to the mobile nodes.

8.3 Accuracy: Identifying Sensing Errors

Thus far we have assumed that the data delivered by the sensor network is simply a subset of the ground truth data. Unfortunately, sensor network data streams are notoriously dirty [22]. Reliably distinguishing errors from correct sensor readings is a difficult problem, as famously illustrated by failure to detect the hole in the ozone layer at the South Pole for six years because the depletion was so severe that it was discarded as erroneous by the software processing the sensor data [25].

Errors in sensor data streams arise from missed data or unreliable readings. Unreliable readings may be either random or systematic. Figure 8.3 shows examples of (from left to right) random noise, systematic bias, and both random error and missing data caused by sensor failure.

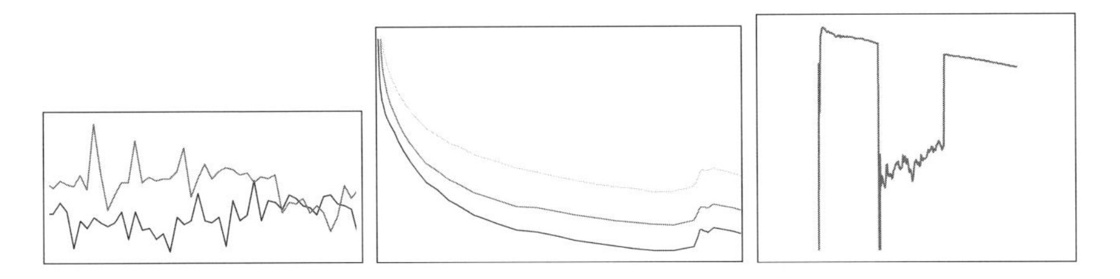

Fig. 8.3 Random, systematic and failure errors in soil moisture sensor readings

Random noise is present in all sensed data, the lowest limit being thermal noise caused by the movement of electrons in the sensing instrument [32]. Random noise may also be introduced by incorrect deployment of a sensor. Sources of systematic errors include sensor miscalibration, low battery power, hostile environment conditions, and errors in the hardware and software used to process the data. Missing readings can be caused by failures at any stage in the chain of events from making a sensor reading, converting it to scientific units, and saving that reading into permanent storage. For example, data may be lost or corrupted during sensing or during radio transmission. Errors may be made when logging data to permanent storage or data may be incorrectly labeled, for example with an incorrect time stamp. The harsh field conditions typical of many sensor network applications only serve to exacerbate errors. For example, clock drift between nodes is significantly worsened in high temperatures in outdoor settings.

The problem of sensing errors in sensory data monitoring is addressed by risk management methods. Risk avoidance is achieved by standard robust system design techniques including thorough testing, hardware and software diversity and redundancy. Risk mitigation is achieved by the application of algorithms for cleaning data streams.

8.3.1 Risk Avoidance for Sensing Errors

Risk is a combined measure of the probability of occurrence of an undesirable event and the severity of adverse effects. Erroneous data are a risk for sensor networks. Systems engineers manage risk by first identifying potential causes of the undesirable event, estimating the probability and severity of that event occurring, and then identifying ways to avoid or mitigate the most significant risks. The severity of an unrecoverable data event is high, since dependable gathering of environmental data is the primary goal of a sensor network. Techniques for reducing the risk of erroneous data rely on redundancy and diversity in the system.

Heterogeneous sensor networks reduce risk by introducing diversity. They use different sensors and different sensing nodes to measure the same phenomena. Networks can include mobile hand-held sensors, traditional loggers, and even physical measuring devices such as rain gauges, as well as wireless sensor nodes. Readings of the same phenomenon from different types of sensors and nodes either improve confidence in the correctness of the data (in the case of agreement) or point to possible errors (in the case of disagreement).

Practical field trials have demonstrated the importance of testing and regression testing of software and hardware both in the laboratory and in the field [12,41]. Effective testing mitigates the risk of software conversion errors, undetected hardware faults, misplacement of sensors in the field, and protocol errors.

Redundancy is most commonly provided by over-sampling the environment using extra sensor nodes and sensors. Sensor network protocols are designed with the assumption that nodes will fail from time to time, and may leave or join the network at different times. When node failure occurs, then nearby nodes will provide similar information. There are, however, several disadvantages of relying on over-sampling including the increased cost and increased volume of data. Also, over-sampling implicitly assumes that the environmental variable being measured varies smoothly. For phenomena such as temperature, this assumption is reasonable at local scale. However, for more dynamic variables such as soil moisture, the assumption is certainly not justified [12].

Another redundancy technique is to log all significant data readings, both at the source and at one or more stages as it is transferred to permanent storage. Logging should be used in conjunction with data compression techniques to avoid overloading local nodes' limited storage.

8.3.2 Risk Mitigation by Data Stream Cleaning

Not all risks of erroneous data can be avoided, since some are the result of circumstances that the sensor network system cannot control. For example, during rain sensor network radios are unreliable, so messages may not be deliverable. In the field, sensors will be damaged, batteries fail, and nodes may be vandalized. In this situation the network will deliver data streams that either contain erroneous readings, or that are missing readings.

Elnahrawy and Nath [15] proposed a cleaning algorithm for data streams using two techniques to reduce uncertainty from random noise. Noise is assumed to be Gaussian with zero mean and variance that is calculated from a number of sources including known noise characteristics of the sensor, knowledge of the observed phenomena, other sensor streams and expert opinion. The algorithm then uses a Bayesian approach and prior knowledge of the true sensor reading to calculate a probability density function (pdf) estimate of the true reading. Another scheme based on prior knowledge was described in [34]. Elnahrawy and Nath also presented an algorithm for evaluating queries over cleaned sensor readings, that is pdfs, for probabilistic queries of the form "return the maximum reading from the set of sensors with at least C% chance that their reading is above threshold T". C is the confidence level required by the end user.

BBQ, a tool to be used with TinyDB, provides end-user support for model-driven data acquisition, in which each phenomena being *measured* is also *modeled* by a pdf [14]. Models are derived from training sets of historical data using standard learning algorithms. BBQ sends queries into the sensor network only in the case that its uncertainty about the reading is too high. Where the model predicts a value with high certainty, then that model value is used. Given a set of models, BBQ's query planning algorithm decide on the optimal way to answer that query. For example, the planner may choose to use a closely correlated, but lower cost reading. For example, battery voltage can be a proxy for temperature, and reading battery voltage has a lower sensor cost and lower energy usage than temperature sensing.

ESP is an extensible framework for cleaning dirty sensor network data [22]. Users instantiate the framework with their own algorithms for any of these steps. ESP provides a set of different data cleaning subtasks, that can be pipelined together by the user as required:

Point: Individual sensor reading tuples are cleaned using user-selected error correction, transformation and filtering algorithms.
Smooth: Aggregate readings over a *temporal* sliding window, to interpolate missing readings and remove single errant readings.
Merge: Aggregate readings over a *spatial* sliding window, to interpolate missing readings and remove single errant readings.
Arbitrate: Conflict resolution between different readings from different sensors.
Virtualize: Cross-checking readings from multiple types of sensors, together with application level cleaning of data.

As well as random noise, sensor network designers need ways to identify and correct systematic bias, or calibration errors, between sensors. Traditional sensing systems rely on micro-calibration, in which there are few sensors in the whole system and each individual sensor is either reliably calibrated in the factory, or have an easy-to-use, built-in calibration interface [45]. However, the sensors used in sensor networks are too numerous for individual calibration procedures and usually too low cost to have reliable, long-term factory calibration. Successful calibration requires both ground truth readings and sensor readings for different parts of the sensor range, as well as a model for the expected behavior of the sensed variable. In general, the problem is to estimate a set of parameters β for each sensor where the calibration function f, the true reading r^* and the actual reading r are all known and $r^* = f(\beta, r)$. Given a set of such equations, least squares can be used to derive β for each sensor. However, obtaining sufficient ground truth data to be able to calibrate sensors in place may not be feasible, and thus the general problem of accurate macro-calibration remains open. The details of algorithms for collaborative macro-calibration in sensor networks are beyond the scope of this chapter, but further details can be found in [45,9].

8.4 Summary

This chapter has presented the state of the art in three areas necessary for effective sensory data monitoring: specifying data, selecting data and cleaning data. A variety of algorithms as well as software and hardware tools have been developed for these tasks.

There have been both successes and significant failures in building and deploying sensor networks [31,40,39,12,10,4,8,43,41]. Perhaps the most important lesson to draw from this experience is the crucial importance of systems engineering principles in designing and building sensor networks. Designers and users must focus on the goals to be achieved, be knowledgeable about the strengths and limitations of the sensor network components used, and follow comprehensive risk management strategies.

Sensory data monitoring is still a field in its infancy, and a number of open research problems remain: Which techniques provide the best interface for users to specify the type of data they are interested in collecting? Which algorithms are most effective for selecting high value data from data streams? Which are the best risk management techniques for preventing errors in data streams and for cleaning data streams that do contain errors?

Acknowledgements The work reported in this chapter was supported by funding from Motorola GSG Perth, CRC Dryland Salinity, and the University of WA. Fleck motes and soil moisture sensor boards used in this research were developed by the Wireless Sensor and Actuator Networks group of CSIRO ICT at the Queensland Centre for Advanced Technologies.

References

[1] A. Adi, O. Etzion, AMIT—the situation manager. The VLDB Journal, 13(2):177–203, 2004.
[2] A. Bakshi, V.K. Prasanna, J. Reich, D. Larner, The abstract task graph: a methodology for architecture-independent programming of networked sensor systems. In: Proceedings of the 2005 Workshop on End-to-End, Sense-and-Respond Systems, Applications and Services (EESR '05), pp. 19–24, Berkeley, CA, USA. USENIX Association, 2005.
[3] M.A. Batalin, M. Rahimi, Y. Yu, D. Liu, A. Kansal, G.S. Sukhatme, W.J. Kaiser, M. Hansen, G.J. Pottie, M. Srivastava, D. Estrin, Call and response: experiments in sampling the environment. In: SenSys '04: Proceedings of the 2nd International Conference on Embedded Networked Sensor Systems, pp. 25–38, New York, NY, USA. ACM Press, 2004.
[4] R. Beckwith, D. Teibel, P. Bowen, Report from the field: results from an agricultural wireless sensor network. In: First IEEE Workshop on Embedded Networked Sensors (EmNets), 2004.
[5] P. Bonnet, J. Gehrke, P. Seshadri, Querying the physical world. IEEE personal communications, 7:10–15, 2000.
[6] A. Boulis, C.-C. Han, M.B. Srivastava, Design and implementation of a framework for programmable and efficient sensor networks. In: The First International Conference on Mobile Systems, Applications, and Services, MobiSys 2003, San Francisco, CA, 2003.
[7] P. Buonadonna, D. Gay, J.M. Hellerstein, W. Hong, S. Madden, TASK: sensor network in a box. In: Proceedings of the Second European Workshop on Wireless Sensor Networks, pp. 133–144, 2005.
[8] J. Burrell, T. Brooke, R. Beckwith, Vineyard computing: sensor networks in agricultural production. IEEE Pervasive Computing 3(1):38–45, 2004.
[9] V. Byckovskiy, S. Megerian, D. Estrin, M. Potkonjak, A collaborative approach to in-place sensor calibration. In: Proceedings of the Second International Workshop on Information

Processing in Sensor Networks (IPSN). Lecture Notes in Computer Science, vol. 2634, pp. 301–316. Springer, Berlin, 2003.
[10] R. Cardell-Oliver, ROPE: a reactive, opportunistic protocol for environment monitoring sensor networks. In: EmNetS-II. The Second IEEE Workshop on Embedded Networked Sensors, Sydney, pp. 63–70, May 2005.
[11] R. Cardell-Oliver, M. Reynolds, M. Kranz, A space and time requirements logic for sensor networks. In: Proceedings of IEEE-ISoLA 2006, to appear, 2007.
[12] R. Cardell-Oliver, K. Smettem, M. Kranz, K. Mayer, A reactive soil moisture sensor network: design and field evaluation. International Journal of Distributed Sensor Networks, 149–162, 2005.
[13] Decagon Echo-20 dielectric aquameter, 2005. [Online] Available at http://www.decagon.com/echo/.
[14] A. Deshpande, C. Guestrin, S.R. Madden, J.M. Hellerstein, W. Hong, Model-based approximate querying in sensor networks. The International Journal on Very Large Data Bases, 417–433, 2005.
[15] E. Elnahrawy, B. Nath, Cleaning and querying noisy sensors. In: WSNA 03: Proceedings of the 2nd ACM International Conference on Wireless Sensor Networks and Applications, New York, NY, USA, pp. 78–87. ACM Press, 2003.
[16] C.-L. Fok, G.-C. Roman, C. Lu, Rapid development and flexible deployment of adaptive wireless sensor network applications. In: Proceedings of the 24th International Conference on Distributed Computing Systems (ICDCS'05), pp. 653–662. IEEE, June 2005.
[17] C. Frank, K. Romer, Algorithms for generic role assignment in wireless sensor networks. In: Proceedings of the 3rd ACM Conference on Embedded Networked Sensor Systems (SenSys'05), November 2005.
[18] R. Gummadi, O. Gnawali, R. Govindan, Macro-programming wireless sensor networks using Kairos. In: Proceedings of the International Conference on Distributed Computing in Sensor Systems (DCOSS), 2005.
[19] G. Hackmann, C.-L. Fok, G.-C. Roman, C. Lu, Agimone: middleware support for seamless integration of sensor and IP networks. In: Lecture Notes in Computer Science, vol. 4026, pp. 101–118, 2006.
[20] J.M. Hellerstein, W. Hong, S. Madden, K. Stanek, Beyond Average: Towards Sophisticated Sensing with Queries. In: 2nd International Workshop on Information Processing in Sensor Networks, IPSN '03, March 2003.
[21] C. Jaikaeo, C. Srisathapornphat, C.-C. Shen, Querying and tasking in sensor networks. In: SPIE's 14th Annual International Symposium on Aerospace/Defense Sensing, Simulation, and Control (Digitization of the Battlespace V), Orlando, Florida, 24–28 April 2000.
[22] S.R. Jeffery, G. Alonso, M.J. Franklin, W. Hong, J. Widom, A pipelined framework for online cleaning of sensor data streams. In: ICDE '06: Proceedings of the 22nd International Conference on Data Engineering (ICDE'06), p. 140, Washington, DC, USA. IEEE Computer Society, 2006.
[23] M. Kranz, SENSID: a situation detector for sensor networks. Honours Thesis, June 2005. Honours Thesis, School of Computer Science and Software Engineering, University of Western Australia.
[24] R. Kumar, M. Wolenetz, B. Agarwalla, J. Shin, P. Hutto, A. Paul, U. Ramachandran, Dfuse: a framework for distributed data fusion. In: Proceedings of the First International Conference on Embedded Networked Sensor Systems (SenSys), pp. 114–125. ACM Press, 2003.
[25] N.G. Leveson, SAFEWARE: System Safety and Computers. Addison–Wesley, Reading, 1995.
[26] P. Levis, D. Culler, Maté: A tiny virtual machine for sensor networks. In: Proceedings of the 10th International Conference on Architectural Support for Programming Languages and Operating Systems (ASPLOS X), 2002.
[27] S. Li, Y. Lin, S.H. Son, J.A. Stankovic, Y. Wei, Event detection using data service middleware in distributed sensor networks. Wireless Sensor Networks of Telecommunications Systems, 26:351–368, 2004.

[28] L. Luo, T.F. Abdelzaher, T. He, J.A. Stankovic, EnviroSuite: an environmentally immersive programming framework for sensor networks. Transactions on Embedded Computing Systems, 5(3):543–576, 2006.
[29] S. Madden, M.J. Franklin, J.M. Hellerstein, W. Hong, TAG: a tiny aggregation service for ad-hoc sensor networks. SIGOPS Operating Systems Review, 36(SI):131–146, 2002.
[30] S. Madden, M.J. Franklin, J.M. Hellerstein, W. Hong, The design of an acquisitional query processor for sensor networks. In: Proceedings of the 2003 ACM SIGMOD International Conference on Management of Data (SIGMOD '03), pp. 491–502, 2003.
[31] A. Mainwaring, J. Polastre, R. Szewczyk, D. Culler, J. Anderson, Wireless sensor networks for habitat monitoring. In: Proc. First ACM International Workshop on Wireless Sensor Networks and Applications, Atlanta, Georgia, USA, September 2002.
[32] J. Meston, Efficiency and robustness in the gathering of data in wireless sensor networks: making every bit count. Honours Thesis, School of Computer Science and Software Engineering, University of Western Australia, November 2005.
[33] M. Moss, Evaluation of event-aware environmental data compression schemes for wireless sensor networks. Honours Thesis, School of Computer Science and Software Engineering, University of Western Australia, November 2005.
[34] D. Mukhopadhyay, S. Panigrahi, S. Dey, Data aware, low cost error correction for wireless sensor networks. In: Wireless Communications and Networking Conference, vol. 4, pp. 2492–2497, March 2004.
[35] C. Rye, Development of a web-based interface for analysis of environmental data from a wireless sensor networks. Honours Thesis, School of Environmental Engineering, University of Western Australia, October 2005.
[36] T. Schoellhammer, B. Greenstein, E. Osterweil, M. Wimbrow, D. Estrin, Lightweight temporal compression of microclimate datasets. In: Proceedings of 29th Annual IEEE International Conference on Local Computer Networks, pp. 516–524. IEEE, 2004.
[37] S. Sen, R. Cardell-Oliver, A rule-based language for programming wireless sensor actuator networks using frequency and communication. In: Proceedings of the third IEEE Workshop on Embedded Networked Sensors (EMNETS'06), June 2006.
[38] C. Srisathapornphat, C. Jaikaeo, C.-C. Shen, Sensor information networking architecture. In: 2000 International Workshop on Parallel Processing, ICPP 2000, Toronto, Canada, August 2000.
[39] R. Szewczyk, A. Mainwaring, J. Polastre, J. Anderson, D. Culler, An analysis of a large scale habitat monitoring application. In: Proceedings of the 2nd International Conference on Embedded Networked Sensor Systems, pp. 214–226. ACM Press, 2004.
[40] R. Szewczyk, J. Polastre, A. Mainwaring, D. Culler, Lessons from a sensor network expedition. In: Proceedings of the First European Workshop on Wireless Sensor Networks, Berlin, Germany, January 2004.
[41] G. Tolle, D. Culler, Design of an application-cooperative management system for wireless sensor networks. In: Second European Workshop on Wireless Sensor Networks (EWSN), Istanbul, Turkey, January 2005.
[42] M. Turon, MOTE-VIEW: a sensor network monitoring and management tool. In: EmNetS-II. The Second IEEE Workshop on Embedded Networked Sensors, Sydney, pp. 11–18, May 2005.
[43] D. Wang, E. Arens, T. Webster, M. Shi, How the number and placement of sensors controlling room air distribution systems affect energy use and comfort. In: International Conference for Enhanced Building Operations, ICEBO 2002, Richardson, Texas, USA, October 2002.
[44] M. Welsh, G. Mainland, Programming sensor networks using abstract regions. In: Proceedings of the First USENIX/ACM Symposium on Networked Systems Design and Implementation, NSDI '04, March 2004.
[45] K. Whitehouse, D. Culler, Macro-calibration in sensor/actuator networks. Mobile Networks and Applications, 8(4):463–472, 2003.
[46] W. Zhenhua, X. Hongbo, T. Yan, T. Jinwen, L.L. Jian, Integer Haar wavelet for remote sensing image compression. In: 6th International Conference on Signal Processing, August 2002.

Part III
Mining Sensor Network Data Streams

Chapter 9
Clustering Techniques in Sensor Networks

Pedro Pereira Rodrigues and João Gama

Abstract The traditional knowledge discovery environment, where data and processing units are centralized in controlled laboratories and servers, is now completely transformed into a web of sensorial devices, some of them with local processing ability. This scenario represents a new knowledge-extraction environment, possibly not completely observable, that is much less controlled by both the human user and a common centralized control process.

9.1 A Ubiquitous Streaming Setting

Clustering is probably the most frequently used data mining algorithm, used as exploratory data analysis or included in other data mining techniques like regression or classification. In recent real-world applications, however, the traditional environment where all data are available all the time is now outdated. Data flow continuously from a data stream at high speed, potentially producing databases with tendentiously infinite length. Moreover, data gathering and analysis has become ubiquitous, in the sense that our world is evolving into a setting where all devices, as small as they may be, will be able to include sensing and processing ability. Thus, if data are to be gathered altogether, the result could be databases with tendentiously infinite width. Hence, new techniques must be defined, or adaptations of known methods should appear, in order to deal with this new ubiquitous streaming setting.

Most works on clustering analysis for sensor networks actually concentrate on clustering the sensors by their geographical position [9] and connectivity, mainly for

P.P. Rodrigues
LIAAD–INESC Porto L.A. and Faculty of Sciences, University of Porto, Rua de Ceuta, 118-6, 4050-190 Porto, Portugal
e-mail: pprodriques@fc.up.pt

J. Gama
LIAAD–INESC Porto L.A. and Faculty of Economics, University of Porto, Rua de Ceuta, 118-6, 4050-190 Porto, Portugal
e-mail: jgama@fep.up.pt

power management [47] and network routing purposes [26]. However, in this chapter, we are interested in clustering techniques for the data produced by the sensors. The motivation for this is all around us. As networks and communications spread out, so does the distribution of novel and advanced measuring sensors. The networks created by this setting can easily include thousands of sensors, each one capable of measuring, analyzing and transmitting data. From another point of view, given the evolution of hardware components, these sensors now act as fast data generators, producing information in a streaming environment.

Given the extent of common sensor networks, the old client-server architecture is practically useless to help the process of clustering data streams produced on sensors. Distributed data mining methods have been proposed such that communication between nodes of the network is enhanced to allow information exchange between sensors.

As an example, consider electricity distribution networks that usually measure load demand on several different sensors. Sensors can act at different granularity levels. For instance, sensors can measure the load of each home in a residential area or the load at substations. The identification of typical load profiles is of great importance for companies' planning purposes. Information of unusual profiles can be used to improving production plans, for example.

Data collected in sensor networks continuously flow for large periods of time, eventually at high speed. Considering their dynamic behavior, clustering over data streams should be addressed as an online and incremental procedure, in order to enable faster adaptation to new concepts and to produce better models over time. Thus, traditional models cannot adapt to the high-speed arrival of new examples, and algorithms developed to deal with this fast scenario usually aim to process data in real time. With respect to clustering analysis, these algorithms should be capable of continuously maintaining a compact description of the most recent data, processing examples in constant time and memory at the rate they arrive.

This chapter introduces a general overview of clustering analysis over data streams produced by sensor networks, focusing on state-of-the-art methods and novel approaches to the problem. We will wander from partitional clustering to hierarchical clustering methods, exploiting the challenges inherent to clustering streaming sensor data.

9.2 The Core of Clustering Procedures

Jain and Dubes [27] defined the clustering procedure as a data organization process which can disclose intrinsic data structure, either as instance groups or a hierarchy of groups. The representation of this process may then be investigated for confirmation of prior data knowledge or motivation for new experiences. However, there's no fixed definition, among the data mining community for the clustering procedure. Nevertheless, Guha et al. [21] presented clustering as: *given n data points in a d-dimensional metric space, partition the data points into k clusters such that the* *data points within a cluster are more similar to each other than data points in different clusters*. The notion of *similarity* is context-specific but, nonetheless, abstract enough to aggregate different problems. This definition already discloses several problems if we consider the application to streaming sensor networks. How does it apply when one does not have n data points but m buckets of n_m data points, spread out on a web of sensors? What happens when these buckets are dynamically filled and emptied by emerging data? Moreover, does it represent the problem of clustering variables over time? These are the emerging questions researchers should address in the dawn of the age of ubiquitous streaming data. Given previous definitions, it is clear that clustering procedure cannot be executed at once. We will pay attention to some inherently relevant components, such as data acquisition, representation, strategy and validation.

9.2.1 Data Acquisition

Traditionally, data used to be acquired from a single source, all at one time. Sometimes, different sources provided data that need to be merged in a single database before being processed. With the advent of data streams, all these data models were forced to change. Now, streaming data are produced by several sources at high speed. Moreover, the speed at which different sources produce data is rarely the same. This way, a centralized process should gather data from the different sources with a non-blocking strategy (e.g. round robin) so that it continuously queries data from the sources, compiling the unique data stream that is going to be analyzed. Distributed data mining systems consider that data should be modeled at each source separately, with synopsis or meta-information being shared among different processes. This is the data-acquisition model that shares more characteristics with the one needed in clustering of streaming sensor networks data.

9.2.2 Data Representation

Prior to any analysis, a rational way to represent data must be defined. Different types of data can usually be found in cluster analysis problems. Anderberg [2] presented a classification of data, according to two features: data *type* and data *scale*; according to data *type*, a single feature may be *binary*, *discrete* or *continuous*; according to data *scale*, variables may have *qualitative* (nominal and ordinal) scales or *quantitative* (interval and ratio) scales [27]. Our focus will be on the analysis of continuous and quantitative variables, as this is the most common data produced by sensors. This restriction allows the research to concentrate on the precise problem, since a difficult step of clustering analysis is skipped, that is, data type and scale determination. Indeed, the clustering procedure can be seen as a process of labeling objects to enable the generation of a nominal scale out of a quantitative one [2].

In order to relate different data points or variables, it is necessary to define the *proximity index* between two individual data examples, which may be of *similarity* such as the *correlation coefficient* or of *dissimilarity* such as the *Euclidean distance*. The more similar two vectors are, the higher their similarity and the lower their distance will be. Many similarity and dissimilarity measures could be used in clustering analysis, and many of them have proved to be outstanding tools for data mining. However, one should only consider the relevance of a proximity index based on the type and scale of the data being analyzed.

9.2.3 Clustering Strategies

What properties must an algorithm possess in order to find groups in data? Unfortunately, the answer to this question isn't at all clear. Some methods aim at finding compact clusters, while others produce spherical clusters. A "*good*" cluster definition may be exceptional following a certain criterion and completely random according to another criterion, being nearly impossible to define what a *good cluster* is. Regardless of the method, Han and Kamber defined several goals for a good clustering procedure, from which we highlight the following as important for clustering sensor networks data [23]:

- Possibility of analysis of heterogeneous attributes;
- Require the least human parameterization as possible, reinforcing the process with high dynamics and low bias;
- Robustness to the order of examples, mainly on methods performing incremental clustering, processing examples once; and
- Provide high levels of interpretation and usability.

Many clustering techniques emerged from research and one can group most of them into two major paradigms: *partitional* [27] and *hierarchical* [29] approaches. Nevertheless, *density* [13,40] and *grid-based* [45,36] systems also present promising research lines. Partitional (or parametric) clustering algorithms are procedures that solve an optimization problem, usually minimizing a cost function, in order to satisfy an optimality criterion imposed by the model. Hierarchical algorithms set their basis on dissimilarities between elements of the same cluster or between clusters, defining a hierarchy of clusters, either in a agglomerative or divisive way. Moreover, they have a major advantage over partitional methods as they do not require user-predefined number of target clusters. An overview of clustering analysis can be found in [23].

9.2.3.1 Clustering Validation

Clustering is usually referred to an unsupervised process [7]. Most of the time, the best way of partitioning the data cannot be determined, nor even the right number of present clusters, resulting in the impossibility of computing an exact evaluation of our model. Nevertheless, the evaluation of the result of a clustering procedure is known as cluster validity [22]. Due to the unsupervised property of the clustering procedure, any validity measure that could be used would only give an indication of the quality of the cluster. Consequently, it can only be considered as a tool, preceding an expert evaluation of the resulting cluster definition. We can generally find three approaches to the investigation of cluster validity [43], usually known as *external*, *internal* and *relative* criteria. For *external* criteria we consider an evaluation based on a pre-specified structure which reflects our prior knowledge or intuition over the data set. An *internal* criterion usually stands for an evaluation based on measures involving the data itself. The third approach compares the resulting cluster structure with other clustering results for the same data set. The results used to make this comparison may be gathered using the same algorithm with different specifications or using different algorithms.

9.3 Clustering Streaming Examples

According to Guha et al. [20], a data stream is an ordered sequence of points that can be read only once or a small number of times. This data model emerged from several new applications, such as sensor data monitoring, web clicks analysis, credit card usage, multimedia data processing, etc., which required a different approach for the usual data mining problems. As the expansion of the Internet continues and ubiquitous computing becomes a reality, we can expect that such data models will become the rule rather than the exception [12]. Moreover, traditional database management systems are not designed to directly support the continuous queries required by these applications [19].

In these applications data continuously flow, eventually at high speed. The processes generating data are non-stationary and dynamic. Clustering in these environments requires a continuous adaptation of the cluster's structure, by incorporating new data at the rate of arrival. In most applications, the cluster structure must reflect the current state of the nature. Change detection and forgetting outdated data are emerging challenges for cluster methods. Having this in mind, a streaming clustering algorithm must observe certain characteristics [5,8]. The algorithm must

- Possess a compact representation;
- Be fast and incremental in processing new examples;
- Execute only one pass (or less) over the data set;
- Have an available response at any time, with information on progress, time left, etc.;
- Be capable of suspension, stop and resume, for incremental processing;
- Have the ability to incorporate new data on a previously defined model;
- Operate inside a limited RAM buffer;

One problem that usually arises with this sort of model is the definition of a minimum number of observations necessary to assure convergence. Techniques based on the Hoeffding bound [24] can be applied to solve this problem, and have in fact been successfully used in online decision trees [12,18,25].

One of the first clustering systems developed, being both incremental and hierarchical, was the *COBWEB*, a conceptual clustering system that executes a hill-climbing search on the space of hierarchical categorization [15]. This method incrementally incorporates objects in a probabilistic categorization tree, where each node is a probabilistic concept representing a class of objects. The gathering of this information is made by means of the categorization process of the object down the tree, updating counts of sufficient statistics while descending the nodes, and executing one of several operations: classify an object according to an existent cluster, create a new cluster, combine two clusters or divide one cluster into several ones.

9.3.1 Partitioning Methods

Bradley et al. [8] proposed the *Single-Pass K-Means*, an algorithm that aims at increasing the capabilities of *k-means* for large data sets. The main idea is to use a buffer where points of the dataset are kept in a compressed way. Extensions to this algorithm appear in [14] and [34]. The *STREAM* [34] system has the goal of minimizing the sum of the squared differences (as in *k-means*) keeping as a restriction the use of available memory. STREAM processes data into batches of m points. These points are stored in a buffer in main memory. After filling the buffer, STREAM clusters the buffer into k clusters. It then summarizes the points in the buffer by retaining only the k centroids along with the number of examples in each cluster. STREAM discards all the points but the centroids weighted by the number of points assigned to it. The buffer is filled in with new points and the clustering process is repeated using all points in the buffer. This approach results in a one-pass, constant-factor approximation algorithm [34]. The main problem is that STREAM never considers data evolution. The resulting clustering can become dominated by the older, outdated data of the stream. An interesting aspect of this algorithm is the ability to compress old information, a relevant issue in data stream processing.

9.3.2 Hierarchical Methods

One major achievement in this area of research was the *Balanced Iterative Reducing and Clustering using Hierarchies* system [48]. The *BIRCH* system builds a hierarchical structure of data, the *CF-tree*, a balanced tree where each node is a tuple (*Clustering Feature*). A clustering feature contains the sufficient statistics for a given cluster: the number of points, the sum of each feature values and the sum of the squares of each feature value. Each cluster feature corresponds to a cluster. They are hierarchically organized in a *CF-tree*. Each non-leaf node in the tree aggregates the information gathered in the children nodes. This algorithm tries to find the best groups with respect to the available memory, while minimizing the amount of input and output. The *CF-tree* grows by aggregation, getting with only one pass over the data a result with complexity $O(N)$. Another use of the *CF-tree* appears in [1]. More than an algorithm, the *CluStream* is a complete system composed by two components, one *online* and another *offline* [1]. Structures called *micro-clusters* are locally kept, having statistical information of data. These structures are defined as a temporal extension of *clustering feature* vectors presented in [48], being kept as images through time, and possessing a pyramidal form. This information is used by the offline component that depends on a variety of user-defined parameters to perform final clustering by an iterative procedure.

The *CURE* system [21], *Clustering Using REpresentatives*, performs a hierarchical procedure that assumes an intermediate approach between centroid based and all-point based techniques. In this method, each cluster is represented by a constant number of points well distributed within the cluster, that capture the extension and shape of the cluster. This process permits the identification of clusters with arbitrary shapes. The *CURE* system also differs from *BIRCH* in the sense that, instead of pre-aggregating all the points, this system gathers a random sample of the data set, using Chernoff bounds [33] in order to obtain the minimum number of examples.

9.3.3 Distributed Clustering on Sensor Networks

If data are being produced in multiple locations, on a wide sensor network, different frameworks could be applied for example clustering on these streams. A first approach consists of a centralized process that would gather data from sensors, analyzing it afterwards in a unique multivariate stream. As previously stated, this model tends to be unapplicable as sensor networks grow unbounded. Another approach could process in two levels, clustering the data produced by each sensor separately, and then compiling the results on a centralized process which would define the final clusters based on the clusters transmitted by each sensor. For example, a strategy of cluster ensembles [42] operate in this way. This approach would, in fact, decrease the amount of data transmitted in the network. However, it still considers a centralized process to define the final clusters. The network could become overloaded if sensors were required to react to the definition of clusters, forcing the server to communicate with all sensors.

The ubiquitous setting created by sensor networks implies different requirements for clustering methods. Given the processing abilities of each sensor, clustering results should be preferably localized on the sensors where this information becomes an asset. Thus, information query and transmission should be considered only on a restricted sensor space, either using flooding-based approaches, where communication is considered only between sensors within a spherical neighborhood of the querier/transmitter, or trajectory-based approaches, where data are transmitted step-by-step on a path of neighbor sensors. A mixture of these approaches is also possible for query re-transmission [39].

Distributed data mining appears to have the necessary features to apply clustering to streaming data [35]. Although few works were directly targeted at data clustering on sensor networks, some distributed techniques are obvious starters in this area of research. A good example of collaborative work that can be done to achieve clustering on sensor networks was developed early in 2001. Kargupta et al. presented a collective principal component analysis (PCA), and its application to distributed cluster analysis [28]. In this algorithm, each node performs PCA, projecting the local data along the principal components, and applies a known clustering algorithm on this projection. Then, each node sends a small set of representative data points to the central site, which performs PCA on this data, computing global principal components. Each site projects its data along the global principal components, which were sent back by the central node to the rest of the network, and applies its clustering algorithm. A description of local clusters is resent to the central site which combines the cluster descriptions using, for example, nearest neighbor methods. This integration of techniques presented good results because selection of representative samples from data clusters is likely to perform better than uniform sampling of data. Other approaches include Klusch et al.'s proposal, a kernel-density-based clustering method over homogeneous distributed data [31]. In fact, this technique does not find a single clustering definition for all the data set. It defines local clustering for each node, based on a global kernel density function, approximated at each node using sampling from signal processing theory. These two techniques possess a good feature, in that they perform only two rounds of data transmission through the network. Other approaches using the K-Means algorithm have been developed for peer-to-peer environments and sensor networks settings [11,4].

Recently, there is rising development of global frameworks that are capable of mining data on distributed sources. Taking into account the lack of resources usually encountered on sensor networks, Resource-Aware Clustering [16] was proposed as a stream clustering algorithm that can adapt to the changing availability of different resources. The system is integrated in a generic framework that enables resource-awareness in streaming computation, monitoring main resources like memory, battery and CPU usage, in order to achieve scalability to distributed sensor networks, by means of adaptation of the algorithm's parameters. Data arrival rate, sampling and number of clusters are examples of parameters that are controlled by this monitoring process.

Learning localized alternative cluster ensembles is a related problem recently targeted by researchers. Wurst et al. developed the LACE algorithm [46], which collaboratively creates a hierarchy of clusters in a distributed way. This approach considered nodes as distributed users, who labeled data according to their own clustering definition, and applied ensemble techniques in order to integrate clusterings provided by different sources, reflecting locality of data, while keeping user-defined clusters. This trade-off between global and local knowledge is now the key point for example clustering procedures over sensor networks.

9.4 Clustering Multiple Data Streams

The task of clustering multiple variables over data streams is not widely studied, so we should start by formally introducing it. Data streams usually consist of variables producing examples continuously over time. The basic idea behind clustering multiple streams is to find groups of variables that behave similarly through time, which is usually measured in terms of distances between the streams.

9.4.1 Motivation

Clustering variables have already been studied in various fields of real-world applications. Many of them, however, could benefit from a data stream approach. For example:

- In electrical supply systems, clustering demand profiles (e.g., industrial or urban) decreases the computational cost of predicting each individual subnetwork load [37].
- In medical systems, clustering medical sensor data (like ECG, EEG, etc.) is useful to determine correlation between signals [41].
- In financial markets, clustering stock prices evolution helps on prevent bankruptcy [32];

All of these problems address data coming from a high-speed stream. Thus, data stream approaches should be considered as possible solutions.

9.4.2 Formal Definition

Let $X = \langle x_1, x_2, \ldots, x_n \rangle$ be the complete set of n data streams and let $X^t = \langle x_1^t, x_2^t, \ldots, x_n^t \rangle$ be the example containing the observations of all streams x_i at the specific time t. The goal of an incremental clustering system for multiple streams is to find (and make available at any time t) a partition P of those streams, where streams in the same cluster tend to be more alike than streams in different clusters. In partitional clustering, searching for k clusters, the result at time t should be a matrix P of $n \times k$ values, where each P_{ij} is one if stream x_i belongs to cluster c_j and zero otherwise.

Specifically, we can inspect the partition of streams in a particular time window from starting time s until current time t, using examples $X^{s..t}$, which would give a temporal characteristic to the partition. In a hierarchical approach to the problem, the same possibilities apply, with the benefit of not having to previously define the target number of clusters, thus creating a structured output of the hierarchy of clusters.

9.4.3 Requirements

The basic requirements usually put forth when clustering examples over data streams are that the system must possess a compact representation of clusters, must process data in a fast and incremental way and should clearly identify changes in the clustering structure [5]. Clustering multiple streams obviously has strong connections to example clustering, so this task shares the same distrusts and, therefore, the same requirements. However, there are some conceptual differences when addressing multiple streams. Nevertheless, systems that aim to cluster multiple streams should:

- Process with constant update time and memory;
- Enable an any time compact representation;
- Include techniques for structural drift detection;
- Enable the incorporation of new, relevant streams; and
- operate with adaptable configuration.

The next sections try to explain the extent to which these features are required to efficiently cluster multiple streams.

9.4.3.1 Constant Update Time and Memory

Given the usual dimensionality of multiple data streams, an exponential or even linear growth in the number of computations with the number of examples would make the system lose its ability to cope with streaming data. Therefore, systems developed to address data streams must constantly update. A perfect setting would be to have a system becoming faster with new examples. Moreover, memory requirements should never depend on the number of examples, as these are tendentiously infinite. From another point of view, when applying variable clustering to data streams, a system could never be supported on total knowledge of available data. Since data are always evolving and multiple passes over them are impossible, all computations should be incrementally conceived. Thus, information is updated continuously, with no increase in memory, and this update requires low time consumption.

9.4.3.2 Any Time Compact Representation

Data streams reveal an issue that imposes the definition of a compact representation of the data used to perform the clustering: it is impossible to store all previously seen data, even if we consider clipping the streams [3]. In example clustering, a usual compact representation of clusters is either the mean or the medoid of the elements associated with that cluster. This way, only a few number of examples are needed to be stored in order to perform comparisons with new data. However, clustering multiple streams is not about comparing new data with old data, but determining and monitoring relations between the streams. Hence, a compact representation must focus on sufficient statistics, used to compute the measures of similarity between the streams, that can be incrementally updated at each new example arrival.

9.4.3.3 Structural Drift Detection

Streams present inherit dynamics in the flow of data that are usually not considered in the concept of usual data mining. The distribution generating the examples of each stream may (and in fact often does) change over time. Thus, new approaches are needed to consider this possibility of change and new methods have been proposed to deal with variable concept drift [17]. However, detecting concept drift as usually conceived for one variable is not the same as detecting concept drift on the clustering structure of several streams [38]. Structural drift is a point in the stream of data where the clustering structure gathered with previous data is no longer valid, since it no longer represents the new relations of proximity and dissimilarity between the streams. Systems that aim at clustering multiple streams should always include methods to detect (and adapt to) these changes in order to maintain an up-to-date definition of the clustering structure through time.

9.4.3.4 Incorporate New Relevant Streams

In current data streams, the number of streams and the number of interesting correlations can be large. However, almost all data mining approaches, especially dealing with streaming data, consider incoming data with fixed width, that is, only the number of examples increase with time. Unfortunately, current problems include an extra difficulty as new streams may be added to the system through time. Given the nature of the task at hand, a clear process of incorporating new streams in a running process must be used, so that the usual growth in data sources is accepted by the clustering system. Likewise, as data sources bloom from all sorts of applications, their importance also fades out as dissemination and redundancy increase, becoming practically irrelevant to the clustering process. A clear identification of these streams should also increase the quality of dissimilarities computed within each cluster.

9.4.3.5 Adaptable Configuration

From the previous requirements, it becomes obvious that the clustering structure and, even more, the number of clusters in the universe of the problem may change over time. This way, approaches with fixed number of target clusters, though still useful in several problems, should be considered only in that precise scope. In general, approaches with adaptable number of target clusters should be favored for the task of clustering multiple streams. Moreover, hierarchical approaches present even more advantages as they inherently conceive a hierarchical relation of sub-clusters, which can be useful for locally detection of changes in the structure.

9.4.4 Compare and Contrast

Clustering multiple streams is in fact an emerging area of research that is closely connected to two other fields: clustering of time series, for its application in the variable domain; and clustering of streaming examples, for its applications to data flowing from high-speed productive streams.

9.4.4.1 Clustering Time Series

Clustering time series can be seen as the *batch* parent of clustering multiple streams, as it embraces the principles of comparing variables instead of examples. However, a lot of research has been done on clustering subsequences of time series instead of whole clustering of the series, which has raised some controversy in the data mining community [30]. Nevertheless, clustering multiple streams approaches whole clustering instead, so most of the existing techniques can be successfully applied, but only if incremental versions are possible. In fact, this is the major drawback of existing whole clustering techniques. Most of the work on clustering time series assumes the series are known in all their extent, failing to cope with infinite number of observations usually inherent to streaming data. Incremental versions of this type of model are good indicators of what can be done to deal with the present problem.

9.4.4.2 Clustering Streaming Examples

Clustering examples on a streaming environment is already widespread in the data mining community as a technique used to discover structures in data over time [5, 20]. This technique presents some proximity to our task, as they both require high-speed processing of examples and compact representation of clusters. Many times, however, example clustering systems use representatives such as means or medoids in order to reduce dimensionality. This reduction is not so straightforward in systems developed to cluster multiple streams, since the reduction in dimensionality based on representatives would have an effect on the variables dimension instead of the examples dimension. Thus, to solve our problem, systems must define and maintain sufficient statistics used to compute (dis)similarities, in an incremental way, which is not an absolute requirement to the previous problem. Therefore, few of the previously proposed models can be adapted to this new task.

9.4.5 Existing Approaches

An area that is close to clustering multiple streams is known as incremental clustering of time series. In fact, this approach complies with almost all features needed to address our problem. Its only restriction is its application to time series, assuming a fixed ordering of the examples. It is not unusual, in streaming environments, for examples to arrive without a specific ordering, creating a new drawback to time series related methods. Nevertheless, we shall inspect the properties of this models, and their application to our task.

9.4.5.1 Composite Correlations

Wang and Wang introduced an efficient method for monitoring composite correlations, i.e., conjunctions of highly correlated pairs of streams among multiple time series [44]. They used a simple mechanism to predict the correlation values of relevant stream pairs at the next time position, using an incremental computation of the correlation, and ranked the stream pairs carefully so that the pairs that are likely to have low correlation values are evaluated first. They showed that the method significantly reduces the total number of pairs for which it is needed to compute the correlation values due to the conjunctive nature of the composites. Nevertheless, this incremental approach has some drawbacks when applied to data streams, and it is not a clustering procedure.

9.4.5.2 Online *K*-Means

Although several versions of incremental, single-pass or online *K-means* may exist, this is a proposal that clearly aims at clustering multiple streams [6]. The basic idea of Online K-Means is that the clusters centers computed at a given time are the initial clusters centers for the next iteration of K-Means. Some of the requirements are clearly met with this approach:

- An efficient preprocessing step is applied, which includes an incremental computation of the distance between data streams, using a Discrete Fourier Transform approximation of the original data; the K-Means procedure applied here is quadratic in the number of clusters, but linear in the number of elements of each block of data used at each iteration.
- At any time, the clusters centers are maintained, and a cluster definition might be extracted as a result.
- As a mean of structural drift detection, the authors proposed a fuzzy approach, which allows a smooth transition among clusters over time.
- The authors did not include procedures to include new streams over time, and this is the main drawback of this approach.
- The algorithm includes a procedure to dynamically update the optimal number of clusters at each iteration, by testing if increasing or decreasing the number of clusters by one unit would produce better results according to a validation index.

This method's efficiency is mainly due to a scalable online transformation of the original data which allows for a fast computation of approximate distances between streams.

9.4.5.3 COD

Clustering On Demand (COD) is a framework for clustering multiple data streams which performs one data scan for online statistics collection and has compact multi-resolution approximations, designed to address the time and the space constraints in a data stream environment [10]. It is divided in two phases: a first online maintenance phase providing an efficient algorithm to maintain summary hierarchies of the data streams and retrieve approximations of the sub-streams; and an offline clustering phase to define clustering structures of multiple streams with adaptive window sizes. Specifically:

- The system encloses an online summarization procedure that updates the statistics in constant time, based on wavelet-based fitting models.
- Statistics are maintained at all times, creating a compact representation of the global similarities, which reduces the responsiveness of the system as it implies a clustering query to extract the active clustering definition.
- The mechanism for detection of change and trends is completely offline, and human-based, which clearly diminishes the adaptability of the procedure.
- The offline characteristic of the clustering process allows the integration of new streams, without much complexity addition, but yet no procedure was introduced to deal with this problem.
- Given the offline query procedure, each query may ask for a different number of k objective clusters; nevertheless, this is a user-defined feature, reducing the adaptability of the whole process.

A good feature of the system is that it allows the user to query with different window sizes (different resolutions) without having to summarize the streams from scratch, and in time linear to the number of streams and data points.

9.4.5.4 ODAC

The Online Divisive-Agglomerative Clustering (ODAC) is an incremental approach for clustering multiple data streams using a hierarchical procedure [38]. It constructs a tree-like hierarchy of clusters of streams, using a top-down strategy based on the correlation between streams. The system also possesses an agglomerative phase to enhance a dynamic behavior capable of structural change detection. The splitting and agglomerative operators are based on the diameters of existing clusters and supported by a significance level given by the Hoeffding bound [24]. Accordingly, we observe that:

- The update time and memory consumption does not depend on the number of examples, as it gathers sufficient statistics to compute the correlations within each cluster; moreover, anytime a split is reported, the system becomes faster as less correlations must be computed.
- The system possesses an any time compact representation, since a binary hierarchy of clusters is available at each time stamp, and does not need to store anything more than the sufficient statistics and the last example to compute the first-order differences.
- An agglomerative phase is included to react to structural changes; these changes are detected by monitoring the diameters of existing clusters.

- This online system was not designed to include new streams along the execution; however, it could be easily extended to cope with this feature.
- Given its hierarchical core, the system possesses a inherently adaptable configuration of clusters.

Overall, this is one of the systems that clearly addresses clustering of multiple streams. It copes with high-speed production of examples and reduced memory requirements, with constant time update. It also presents adaptability to new data, detecting and reacting to structural drift.

9.5 Open Issues on Clustering Sensor Data Streams

As previously discussed, clustering multiple data streams has already been targeted by researchers seeking ways to cope with the high-speed production of data. However, if these data are produced by sensors on a wide network, the proposed algorithms tend to deal with them as a centralized multivariate stream, without taking into account the locality of data, the transmission and processing resources of sensors, and the breach in the transmitted data quality. Moreover, these algorithms tend to be designed as a single process of analysis without the distributed feature of already developed example clustering systems. Distributed implementations of well-known algorithms may produce both valuable and impractical systems, so the path to them should be carefully inspected.

Considering the main restrictions of sensor networks, the analysis of clusters of multiple sensor streams should comply not only with the requirements for clustering multiple data streams but also with the available resources and setting of the corresponding sensor network. For example, considering the previous example of electricity distribution networks, if a distributed algorithm for clustering streaming sensors is integrated on each sensor, how can local nodes process data and the network interact in order to cluster similar behaviors produced by sensors far from each other, without a centralized monitoring process? How many hops should the network need to perform that analysis? What is the relevance of this information? And what is the relation of this information with the geographical location of sensors?

Future research developments are requested to address these issues, and surely researchers will focus on distributed data mining utilities for large sensor networks streaming data analysis, as sensors and their respective data become more and more ubiquitous and embedded in everyday life.

Acknowledgements Pedro P. Rodriques is supported by a PhD grant attributed by the Portuguese Science and Technology Foundation (SFRH/BD/29219/2006). The authors also wish to acknowledge the participation of projects RETINAE (PRIME/IDEIA/70/00078) and ALES II (POSC/EIA/55340/2004).

References

[1] C.C. Aggarwal, J. Han, J. Wang, P.S. Yu, A framework for clustering evolving data streams. In: VLDB 2003, Proceedings of 29th International Conference on Very Large Data Bases, pp. 81–92. Morgan Kaufmann, September 2003.
[2] M.R. Anderberg, Cluster Analysis for Applications. Academic Press, San Diego, 1973.
[3] A. Bagnall, G. Janacek, Clustering time series with clipped data. Machine Learning, 58(2–3):151–178, 2005.
[4] S. Bandyopadhyay, C. Giannella, U. Maulik, H. Kargupta, K. Liu, S. Datta, Clustering distributed data streams in peer-to-peer environments. Information Sciences, 176(14):1952–1985, 2006.
[5] D. Barbará, Requirements for clustering data streams. SIGKDD Explorations, 3(2):23–27, 2002. Special Issue on Online, Interactive, and Anytime Data Mining.
[6] J. Beringer, E. Hüllermeier, Online clustering of parallel data streams. Data and Knowledge Engineering, 58(2):180–204, 2006.
[7] M.J.A. Berry, G. Linoff, Data Mining Techniques: For Marketing, Sales and Customer Support. Wiley, USA, 1996.
[8] P. Bradley, U. Fayyad, C. Reina, Scaling clustering algorithms to large databases. In: Proceedings of the Fourth International Conference on Knowledge Discovery and Data Mining, pp. 9–15. AAAI Press, 1998.
[9] H. Chan, M. Luk, A. Perrig, Using clustering information for sensor network localization. In: First IEEE International Conference on Distributed Computing in Sensor Systems, pp. 109–125, 2005.
[10] B.-R. Dai, J.-W. Huang, M.-Y. Yeh, M.-S. Chen, Adaptive clustering for multiple evolving streams. IEEE Transactions on Knowledge and Data Engineering, 18(9):1166–1180, 2006.
[11] S. Datta, K. Bhaduri, C. Giannella, R. Wolff, H. Kargupta, Distributed data mining in peer-to-peer networks. IEEE Internet Computing, 10(4):18–26, 2006.
[12] P. Domingos, G. Hulten, Mining high-speed data streams. In: Proceedings of the Sixth ACM-SIGKDD International Conference on Knowledge Discovery and Data Mining, pp. 71–80, Boston, MA, ACM Press, 2000.
[13] M. Ester, H.-P. Kriegel, J. Sander, X. Xu, A density-based algorithm for discovering clusters in large spatial databases with noise. In: E. Simoudis, J. Han, U. Fayyad (Eds.) Second International Conference on Knowledge Discovery and Data Mining, pp. 226–231, Portland, Oregon. AAAI Press, 1996.
[14] F. Farnstrom, J. Lewis, C. Elkan, Scalability for clustering algorithms revisited. SIGKDD Explorations, 2(1):51–57, 2000.
[15] D.H. Fisher, Knowledge acquisition via incremental conceptual clustering. Machine Learning, 2(2):139–172, 1987.
[16] M.M. Gaber, P.S. Yu, A framework for resource-aware knowledge discovery in data streams: a holistic approach with its application to clustering. In: Proceedings of the ACM Symposium on Applied Computing, pp. 649–656, 2006.
[17] J. Gama, P. Medas, G. Castillo, P. Rodrigues, Learning with drift detection. In: A.L.C. Bazzan, S. Labidi (Eds.) Advances in Artificial Intelligence—SBIA 2004, Lecture Notes in Computer Science, vol. 3171, pp. 286–295. Springer, Berlin, 2004.
[18] J. Gama, P. Medas, P. Rodrigues, Learning decision trees from dynamic data streams. In: H. Haddad, L.M. Liebrock, A. Omicini, R.L. Wainwright (Eds.) SAC '05: Proceedings of the 2005 ACM Symposium on Applied Computing, pp. 573–577. ACM Press, 2005.
[19] J. Gama, P. Rodrigues, J. Aguilar-Ruiz, An overview on learning from data streams. New Generation Computing, 25(1):1–4, 2007.
[20] S. Guha, A. Meyerson, N. Mishra, R. Motwani, L. O'Callaghan, Clustering data streams: Theory and practice. IEEE Transactions on Knowledge and Data Engineering, 15(3):515–528, 2003.

[21] S. Guha, R. Rastogi, K. Shim, CURE: An efficient clustering algorithm for large databases. In: L.M. Haas, A. Tiwary (Eds.) Proceedings of the 1998 ACM-SIGMOD International Conference on Management of Data, pp. 73–84. ACM Press, 1998.
[22] M. Halkidi, Y. Batistakis, M. Varzirgiannis, On clustering validation techniques. Journal of Intelligent Information Systems, 17(2–3):107–145, 2001.
[23] J. Han, M. Kamber, Data Mining: Concepts and Techniques. Morgan Kaufmann, San Mateo, 2001.
[24] W. Hoeffding, Probability inequalities for sums of bounded random variables. Journal of the American Statistical Association, 58(301):13–30, 1963.
[25] G. Hulten, L. Spencer, P. Domingos, Mining time-changing data streams. In: Proceedings of the Seventh ACM SIGKDD International Conference on Knowledge Discovery and Data Mining, pp. 97–106. ACM Press, 2001.
[26] J. Ibriq, I. Mahgoub, Cluster-based routing in wireless sensor networks: issues and challenges. In: International Symposium on Performance Evaluation of Computer and Telecommunication Systems, pp. 759–766, 2004.
[27] A.K. Jain, R.C. Dubes, Algorithms for Clustering Data. Prentice Hall, New York, 1988.
[28] H. Kargupta, W. Huang, K. Sivakumar, E.L. Johnson, Distributed clustering using collective principal component analysis. Knowledge and Information Systems, 3(4):422–448, 2001.
[29] L. Kaufman, P.J. Rousseeuw, Finding Groups in Data: An Introduction to Cluster Analysis. Wiley, New York, 1990.
[30] E.J. Keogh, J. Lin, W. Truppel, Clustering of time series subsequences is meaningless: Implications for previous and future research. In: Proceedings of the IEEE International Conference on Data Mining, pp. 115–122. IEEE Computer Society Press, 2003.
[31] M. Klusch, S. Lodi, G. Moro, Distributed clustering based on sampling local density estimates. In: Proceedings of the International Joint Conference on Artificial Intelligence, pp. 485–490, 2003.
[32] R.N. Mantegna, Hierarchical structure in financial markets. The European Physical Journal B, 11(1):193–197, 1999.
[33] R. Motwani, P. Raghavan, Randomized Algorithms. Cambridge University Press, Cambridge, 1995.
[34] L. O'Callaghan, A. Meyerson, R. Motwani, N. Mishra, S. Guha, Streaming-data algorithms for high-quality clustering. In: Proceedings of the Eighteenth Annual IEEE International Conference on Data Engineering, pp. 685–696. IEEE Computer Society, 2002.
[35] B. Park, H. Kargupta, Distributed data mining: algorithms, systems, and applications. In: N. Ye (Ed.) Data Mining Handbook, pp. 341–358. IEA, 2002.
[36] N.H. Park, W.S. Lee, Statistical grid-based clustering over data streams. SIGMOD Record, 33(1):32–37, 2004.
[37] P.P. Rodrigues, J. Gama, Online prediction of clustered streams. In: J. Gama, J. Aguilar-Ruiz, R. Klinkenberg (Eds.) Proceedings of the Fourth International Workshop on Knowledge Discovery from Data Streams, pp. 23–32. ECML/PKDD, 2006.
[38] P. Pereira Rodrigues, J. Gama, J.P. Pedroso, ODAC: Hierarchical clustering of time series data streams. In: J. Ghosh, D. Lambert, D. Skillicorn, and J. Srivastava (Eds.) Proceedings of the Sixth SIAM International Conference on Data Mining, pp. 499–503. SIAM, 2006.
[39] N. Sadagopan, B. Krishnamachari, A. Helmy, Active query forwarding in sensor networks. Ad Hoc Networks, 3(1):91–113, 2005.
[40] G. Sheikholeslami, S. Chatterjee, A. Zhang, WaveCluster: a multi-resolution clustering approach for very large spatial databases. In: Proceedings of the 24th International Conference on Very Large Data Bases, VLDB, pp. 428–439. New York, USA. ACM Press, 1998.
[41] D.M. Sherrill, M.L. Moy, J.J. Reilly, P. Bonato, Using hierarchical clustering methods to classify motor activities of COPD patients from wearable sensor data. Journal of Neuroengineering and Rehabilitation, 2(16), 2005.
[42] A. Strehl, J. Ghosh, Cluster ensembles—a knowledge reuse framework for combining multiple partitions. Journal of Machine Learning Research, 3:583–617, 2002.
[43] S. Theodoridis, K. Koutroumbas, Pattern Recognition. Academic Press, San Diego, 1999.

[44] M. Wang, X.S. Wang, Efficient evaluation of composite correlations for streaming time series. In: Advances in Web-Age Information Management—WAIM 2003, Lecture Notes in Computer Science, vol. 2762, pp. 369–380, Chengdu, China, August 2003. Springer, Berlin, 2003.

[45] W. Wang, J. Yang, R.R. Muntz, STING: A statistical information grid approach to spatial data mining. In: M. Jarke, M.J. Carey, K.R. Dittrich, F.H. Lochovsky, P. Loucopoulos, M.A. Jeusfeld (Eds.) Proceedings of the Twenty-Third International Conference on Very Large Data Bases, pp. 186–195, Athens, Greece. Morgan Kaufmann, 1997.

[46] M. Wurst, K. Morik, I. Mierswa, Localized alternative cluster ensembles for collaborative structuring. In: Proceedings of the 17th European Conference on Machine Learning, Lecture Notes in Computer Science, vol. 4212, pp. 485–496. Springer, Berlin, 2006.

[47] O. Younis, S. Fahmy, HEED: A hybrid, energy-efficient, distributed clustering approach for ad hoc sensor networks. IEEE Transactions on Mobile Computing, 3(4):366–379, 2004.

[48] T. Zhang, R. Ramakrishnan, M. Livny, BIRCH: An efficient data clustering method for very large databases. In: Proceedings of the 1996 ACM SIGMOD International Conference on Management of Data, pp. 103–114. ACM Press, 1996.

Chapter 10
Predictive Learning in Sensor Networks

João Gama and Rasmus Ulslev Pedersen

Abstract Sensor networks act in dynamic environments with distributed sources of continuous data and computing with resource constraints. Learning in these environments is faced with new challenges: the need to continuously maintain a decision model consistent with the most recent data. Desirable properties of learning algorithms include: the ability to maintain an any time model; the ability to modify the decision model whenever new information is available; the ability to forget outdated information; and the ability to detect and react to changes in the underlying process generating data, monitoring the learning process and managing the trade-off between the cost of updating a model and the benefits in performance gains. In this chapter we illustrate these ideas in two learning scenarios—centralized and distributed—and present illustrative algorithms for these contexts.

10.1 Introduction

Most machine learning and data mining approaches assume that examples are independent, identically distributed and generated from a stationary distribution. In that context, standard data mining techniques use finite training sets and generate static models. In contrast, sensor networks are geographically distributed and produce high-speed distributed data streams. They act in dynamic environments and are influenced by adversary conditions. Data collected from a sensor network are correlated, and the distribution is non-stationary. From a data mining perspective, sensor network problems are characterized by a large number of variables (sensors), producing a continuous flow of data, in a dynamic non-stationary environment. One important problem is predicting the measurement of interest of a sensor for a given time horizon.

J. Gama
LIAAD, University of Porto, R. de Ceuta, 118-6, 4050-190 Porto, Portugal
e-mail: jgama@fep.up.pt

R.U. Pedersen
Department of Informatics, Copenhagen Business School, Copenhagen, Denmark
e-mail: rup.inf@cbs.dk

We can consider two different approaches:

- *Centralized Approach.* Sensors send data to a central server which runs the mining algorithms. This is the simplest solution. Its main problem is that the network's dimensionality can be very large.
- *Distributed Approach.* Mining algorithms run in the sensor itself. Sensors are smart devices, although with limited computational power, that can detect and communicate with neighbors. Data mining in this context becomes ubiquitous and distributed.

A horizontal property of the both approaches is the continuous flow of data. Learning from continuous sources of data is faced with new challenges: we need to continuously maintain a decision model consistent with the most recent data. In the next section we discuss the general issues and key properties in learning from data streams. Section 10.3 presents the most representative learning algorithms able to process data streams. Section 10.4 presents kernel-based algorithms for distributed approaches. The last section discusses emerging challenges when learning from data streams.

10.2 General Issues

Domingos and Hulten [24] identified desirable properties of learning systems that are able to mine continuous, high-volume, open-ended data streams as they arrive. According to the authors, these systems should:

- Require small constant time per data example;
- Use a fixed amount of main memory, irrespective of the total number of examples;
- Build a decision model using a single scan over the training data;
- Possess an any time model;
- Be independent from the order of the examples; and
- Be able to deal with concept drift. For stationary data, the system be able to produce decision models that are nearly identical to the ones we would obtain using a batch learner.

These ideas were the basis for the *Very Fast Machine Learning* toolkit.[1] VFML contains tools for learning decision trees, for learning the structure of belief nets (aka Bayesian networks), and for clustering.

Desirable properties of learning algorithms include: an ability to maintain an any time model; an ability to modify the decision model whenever new information is available; the ability to forget outdated information; an ability to detect and react to changes in the underlying process generating data; and the capacity to monitor the learning process and manage the trade-off between the cost of updating a model and the benefits in performance gains.

[1] http://www.cs.washington.edu/dm/vfml/.

Mining in the context of sensor networks must avoid the highly unprovable assumption that the examples are generated at random according to a stationary probability distribution. At least in complex systems and for large time periods, we should expect changes in the distribution of the examples. A natural approach for these *incremental tasks* are *adaptive learning algorithms*, incremental learning algorithms that take into account concept drift.

10.2.1 Incremental Issues

Solutions to these problems require new sampling and randomizing techniques, and new approximate and incremental algorithms. Some data stream models allow delete and update operators. For these models or in the presence of context change, the incremental property is not enough. Learning algorithms need forgetting operators that reverse learning: decremental unlearning [6].

The incremental and decremental issues require permanent maintenance and updating of the decision model as new data are available. Of course, there is a trade-off between the cost of the update and any gain in performance. The update of the model depends on the complexity of its representation language. Very simple models, using few free parameters, can be quite efficient in variance management, and effective in incremental and decremental operations (for example naive Bayes) being a natural choice in the sliding windows framework. The main problem with simple approaches is the boundary in generalization performance they can achieve, since they are limited by high bias. Large volumes of data require efficient bias management. Complex tasks requiring more complex models increase the search space and the cost for structural updating. These models require efficient control strategies for the trade-off between the gain in performance and the cost of updating.

An illustrative example is the case of multiple models. Theoretical results show that it is possible to obtain arbitrary low errors by increasing the number of models. The behavior is exponential. To achieve a linear reduction of the error, we need an exponential increase in the number of models. Find the *breakeven* point, that is the point where costs equal benefits, and going beyond that is a main challenge.

10.2.2 Change Detection Issues

When data flow over time, and at least for large periods of time, the assumption that the examples are generated at random according to a stationary probability distribution is highly unprovable. At least in complex systems and for large time periods, we should expect changes in the distribution of the examples. Learning algorithms must be able not only to incorporate new information when it is available, but also to *forget* old information when it becomes outdated.

Concept drift means that the concept related to the data being collected may shift from time to time, each time after some minimum permanence. The evidence for changes in a concept are reflected in some way in the training examples. Old observations, that reflect past behavior, become irrelevant to the current state of the phenomena under observation and the learning agent must forget that information.

The nature of change is diverse. Changes may occur in the context of learning, due to changes in hidden variables, or in the characteristic properties of the observed variables. Most learning algorithms use blind methods that adapt the decision model at regular intervals without considering whether changes have really occurred. Much more interesting is explicit change-detection mechanisms. The advantage is that they can provide a meaningful description (indicating change-points or small time windows where the change occurs) and quantification of the changes. They may follow two different approaches:

1. Monitoring the evolution of performance indicators through adapting techniques used in statistical process control [19,30].
2. Monitoring distributions on two different time windows. Most of the methods in this approach monitor the evolution of a distance function between two distributions: from past data in a *reference window* and in a current window of the most recent data points [29].

The main research issue is how to incorporate change detection mechanisms in the learning process. Embedding change detection methods in the learning algorithm is a requirement in the context of continuous flow of data. The level of *granularity* of decision models is a relevant property, because it can allow partial, fast and efficient updates in the decision model instead of rebuilding a complete new model whenever a change is detected. The ability to recognize seasonal and re-occurring patterns are open issues.

10.2.3 Evaluation Issues

A key point in any intelligent system is the evaluation methodology. Learning systems generate compact representations of what is being observed. They should be able to improve with experience and continuously self-modify their internal state. Their representation of the world is approximate. How approximate is the representation of the world?

Evaluation is used in two contexts: inside the learning system to assess hypotheses, and as a wrapper over the learning system to estimate the applicability of a particular algorithm in a given problem.

10.2.3.1 Evaluation Metrics

For predictive learning tasks (classification, and regression) the learning goal is to induce a function $y = \hat{f}(\mathbf{x})$. The most relevant dimension is the *generalization error*. It is an estimator of the difference between $\hat{f}$ and the unknown f, and an estimate of the loss that can be expected when applying the model to future examples. The predictions generated by the system can lead to actions with certain costs and benefits (utility, payoff). In many domains the errors have unequal costs. Examples of such applications include credit, marketing, fraud detection, etc.

In batch learning, cross-validation and variants are the standard method to evaluate learning systems using finite training sets. Cross-validation is appropriate for restricted size data sets, generated by stationary distributions, and assuming the independence of examples. In the data streams context, where data are potentially infinite, and generated by time-changing distributions, we need other evaluation strategies. Two viable alternatives are:

1. Train-Test. Hold out an independent test set. Given large enough training and test sets, the algorithm learns a model from the training set and makes predictions for test set examples.
2. Predictive Sequential. *Prequential* [11], where the error of a model is computed from the sequence of examples. For each example the actual model makes a prediction based only on the example attribute values. The prequential-error is computed based on an accumulated sum of a loss function between the prediction and observed values.

The prequential approach provides much more information. It gives us the evolution of the error, a type of learning curve. Nevertheless, it is affected by the order of the examples. In any case, few works study the evaluation component when learning from data streams. This is an important aspect reserved for future work.

10.2.3.2 Evaluation Methodology in Drift Environments

Changes occur over time. Drift detection algorithms assume that data are sequential. Standard evaluation methods, like cross-validation, assume examples are independent and identically distributed. How to evaluate learning algorithms in sequential data? The main constraint is that we must guarantee all test examples have a time stamp larger than those in the training data. Any sampling method that satisfies this constraint can be used.

What should we evaluate? Some key points include:

- Probability of false alarms.
- Probability of change detection.
- Detection rate, defined as the number of examples required to detect a change after the occurrence of a change.
- Evolution of decision model error during the learning process.

10.2.4 Algorithmic Design Choices for Distributed Kernel Learning

In addition to the issues already raised, it is beneficial to look at a concrete example of some choices that arise when one has to design distributed kernel learners. We have chosen to divide the choices into three main categories: kernels, algorithms and models. It is our belief that this set of considerations covers many of the most important decisions to be made in the design and deployment process for distributed learning. The scope of our analysis is limited to an algorithm in which at least the kernel plays a significant role and usually we will focus on the support vector approach. In order to avoid a trivial repetition of known algorithms, we cast each of the analyzed algorithms in a *co-active* setting. For keeping the principles and core ideas as clear as possible, we shall restrict the introduction of *co-active* versions of some single-node algorithms to a dual-node setting. The extension to a higher number of nodes should be similar. It should be noted that game theory is a possible direction to address the co-active learning schemes. To make the overview more useful we use the support vector machine as one common example in the next sections.

10.2.4.1 Classification

Classification is the simplest (and by far the most used) task of separating two classes from one another. The two-class problem is called a binary classification problem. In Fig. 10.1 two SVMs exchange information for a classification example by sending the support vectors from one SVM to the other.

Fig. 10.1 Coactive classification

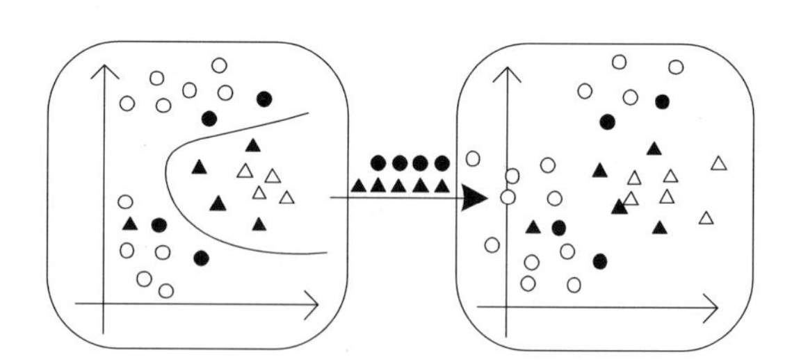

10.2.4.2 Novelty Detection

The co-active novelty detection is between nodes that essentially perform a one-class problem. The sensor's task is to determine if there is a state change in the data it observes. Another aspect of this task is outlier detection, which is to determine if a new observation deviates from previously observed observations. The training phase of the novelty detection algorithm is performed either centrally, and then the node is loaded with the support vectors, or is done on the node and it can use some threshold value to determine if the values are novel (same as an outlier) and fire on the event by passing either a binary signal throughout the network or transmitting the value to the gateway sensor. The co-active mode is obtained by making the sensors exchange support vectors, which are the examples that define the border of the decision support line. The SVMs in Fig. 10.2 exchange the novel points to learn from one another.

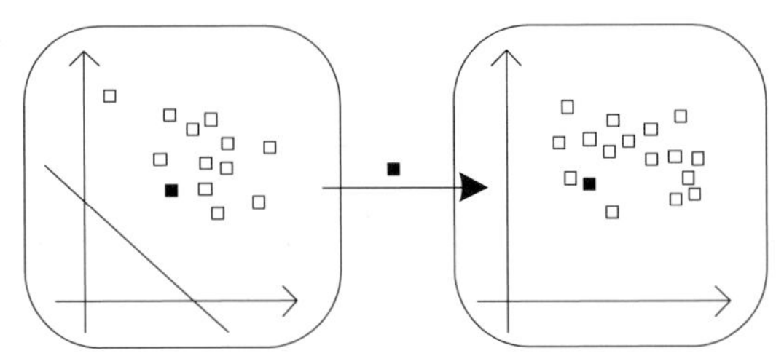

Fig. 10.2 Coactive novelty detection

10.2.4.3 Regression

The regression problems are characterized by the real-valued dependent value that is being predicted by the model. A regression problem takes a number of input dimensions and seeks a model that optimally predicts a real value associated with the particular input vector. The optimal solution is reached by minimizing a particular loss function. The most widely used method for this is to minimize the sum of the squared errors (a squared error is the squared distance from the prediction and the actual output). The SVM loss function is constructed differently, which indirectly turns out to be an advantage for the co-active regression problem. The reason is that data points that are within a distance ϵ from the regression line do not become part of the SVM regression model. This turns out to provide a way to avoid transmitting all but the support vectors to other co-active sensors. It is exemplified in Fig. 10.3.

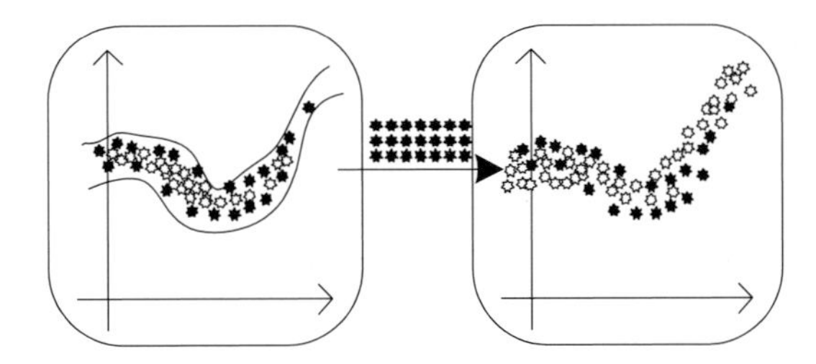

Fig. 10.3 Coactive regression

10.2.4.4 Ranking

Ranking problems arise where the relative importance of the data points are the important issue, as opposed to regression where the difference in the real output is what determines how different two data points are. In ranking problems, we use what are called ordinal variables that provide a relative difference between data points. The ranking example with SVMs follow the idea of the previous figures and it is shown in Fig. 10.4.

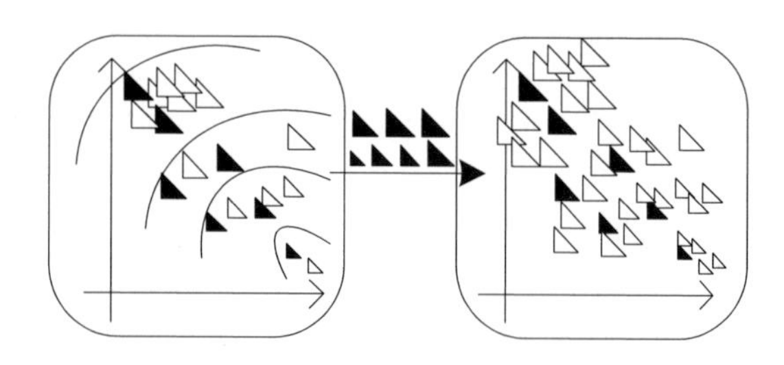

Fig. 10.4 Coactive ranking

10.2.5 KAM: Co-active Modes

We introduce the term *mode* to symbolize the choices available to the inference model designer. In the previous section, we primarily focused on the co-active mode as it was the only available choice. However, this mode can be sub-categorized into batch-, online- and query-based learning. Each of these categories open up for even more detailed design choices if the models can be used by the co-active sensors. Batch learning is the situation where the model is trained in full on the whole data set. Online learning focuses on models where the model is allowed to constantly re-train each time it is fed new data.

10.2.5.1 Batch

The batch mode of training is when the model is static with respect to a training set of data. Before deployment, the model is created using the available data to set its parameters. An advantage of batch training and SVMs is that a large percentage of the data can be retired immediately after training is performed. Namely, the non-support vectors can be retired from memory following training.

10.2.5.2 Streams

This mode of training is batch training with constant re-training as new data arrive at the sensor node. In addition, the time information related to the data is kept and used in some data-retirement scheme. It can be a time stamp associated with the data or simply a FIFO (first in, first out) queue of data. The SVM can be used in various ways to ensure a retirement scheme of data which are believed not to be part of a current state of the system. An implicit manner to restrict the influence of certain (old) data is to restrict the size that the corresponding Lagrange multiplier can take.

If the SVM system was generally trained with $C = 10$, then the influence of older data can be reduced by working with individual lower values of C for some data. It is easy to retire old data in a binary classification SVM model, as the condition $\sum y\alpha$ is directly updateable. If a new data item is inserted into the model as an old one is removed, then the new data item can take on the α value of the old item if the class number (± 1) is the same as the retired item. One way to imagine an online learning scheme is if the sensor network receives a mix of classified and non-classified items. Then the classified items are used on the local SVM models and exchanged among the sensors to provide a co-active learning scheme. The constant feed of non-classified items is then subject to a classification on the nodes. The SVM can be constructed such that it is incrementally trained when new data points arrive. Data points can similarly be unlearned [7]. The LASVM has also been noted to give good results on active learning [2].

10.2.6 Query

The query model mode provides some advantages and disadvantages to the other learning modes. First, we will define how query learning works for SVMs. The model is trained on classified data, but it also receives unclassified data and if this data could be classified then the model would benefit from it. There is a cost of classifying an unlabeled data point and using it in training. This query is submitted to the network and the other node(s) would classify the data point and return the label. It is possible to create a multitude of problems around this scheme with one example being majority voting among the other nodes. This scheme also addresses another important subject which is related to data confidentiality. If some nodes have data which cannot be shared with other nodes, it is still possible for other nodes to query the restricted node and get access to the knowledge. Due to the non-linear nature of the SVM problem space, it would generally not be particularly easy to infer the underlying data points of the query-only nodes. For example, the nodes can restrict the number of queries from any single sensor to ensure that a grid-like search is not conducted to infer the full model.

10.3 Centralized Approaches

10.3.1 Decision Trees

Decision trees are one of the most studied methods in knowledge discovery from data. The most promising research line starts from the observation that a small number of examples are enough to select the correct splitting test and expand a leaf. In this approach, we only need to maintain a set of sufficient statistics at each decision node. The algorithm only makes a decision, that is it installs a split-test at that node when there is enough statistical evidence in favor of a particular split-test. This is the case of [23,13,21]. A notable example is the VFDT system [13]. It can manage millions of examples using few computational resources with a performance similar to a batch decision tree given enough examples.

10.3.1.1 The *Very Fast Decision Tree* Algorithm

The main problem in decision tree induction is the choice of the splitting test to install in each decision node. The VFDT algorithm is based on the simple idea that *a small sample of examples is enough to choose a splitting test*. The main contribution of VFDT is the sound method for defining the size of sample to achieve, with a probability so high as desired, the correct choice.

In VFDT a decision tree is learned by recursively replacing leaves with decision nodes. Each leaf stores the sufficient statistics about attribute values. The sufficient statistics are those needed by a heuristic evaluation function that evaluates the merit of split tests based on attribute values. When an example is available, it traverses the tree from the root to a leaf, evaluating the appropriate attribute at each node, and following the branch corresponding to the attribute's value in the example. When the example reaches a leaf, the sufficient statistics are updated. Then, each possible condition based on attribute values is evaluated. If there is enough statistical support in favor of one test over the others, the leaf is changed to a decision node. The new decision node will have as many descendant leaves as the number of possible values for the chosen attribute (therefore this tree is not necessarily binary). The decision nodes only maintain the information about the split test installed in this node.

The main innovation of the VFDT system is the use of Hoeffding bounds to decide how many examples are necessary to observe before installing a split test at each leaf. Suppose we have made n independent observations of a random variable r whose range is R. The Hoeffding bound states, with probability $1-\delta$, that the true average of r, $\bar{r}$, is at least $\bar{r} - \epsilon$ and $\epsilon = \sqrt{R^2 \frac{\ln(\frac{1}{\delta})}{2n}}$.

Let $H(\cdot)$ be the evaluation function of an attribute. For the information gain, the range R, of $H(\cdot)$ is $\log_2(k)$ where k is denotes the number of classes. Let x_a be the attribute with the highest $H(\cdot)$, x_b the attribute with second highest $H(\cdot)$ and $\overline{\Delta H} = \overline{H}(x_a) - \overline{H}(x_b)$, the difference between the two better attributes. Then if $\overline{\Delta H} > \epsilon$ with n examples observed in the leaf, the Hoeffding bound states with probability $1 - \delta$ that x_a is really the attribute with highest value in the evaluation function. In this case the leaf must be transformed into a decision node that splits on x_a.

The evaluation of the merit function for each example could be very expensive. It turns out that it is not efficient to compute $H(\cdot)$ every time an example arrives. VFDT only computes the attribute evaluation function $H(\cdot)$ when a minimum number of examples has been observed since the last evaluation. This minimum number of examples is a user-defined parameter. When two or more attributes continuously have very similar values of $H(\cdot)$, even with a large number of examples, the Hoeffding bound will not decide between them. To solve this problem the VFDT uses

Algorithm 1: The Hoeffding Tree Algorithm

input : S: A Sequence of Examples
X: A Set of nominal Attributes
Y: $Y = \{y_1, \ldots, y_k\}$ Set of Class values
$H(.)$: Split Evaluation Function
δ: is one minus the desired probability
of choosing the correct attribute at any node.
τ: Constant to solve ties.
output: HT: is a Decision Tree
begin
  Let $HT \leftarrow$ Empty Leaf (Root)
  foreach $y_k \in Y$, $x_i \in X$, *value j of* x_i **do**
    $x_{ijk} \leftarrow 0$
  foreach *example* $(x, y_k) \in S$ **do**
    Traverse the tree HT from the root till a leaf l
    foreach *value j of* x_i **do**
      Increment x_{ijk}
    if *all examples in l are not of the same class* **then**
      Compute $G_l(X_i)$ for all the attributes
      Let X_a be the attribute with highest H_l
      Let X_b be the attribute with second highest H_l
      Compute ϵ (Hoeffding bound)
      if $(H(X_a) - H(X_b) > \epsilon)$ **then**
        Replace l with a splitting test based on attribute X_a
        Add a new empty leaf for each branch of the split
      else
        if $\epsilon < \tau$ **then**
          Replace l with a splitting test based on attribute X_a
          Add a new empty leaf for each branch of the split
end

a constant τ introduced by the user for run-off, e.g., if $\overline{\Delta H} < \epsilon < \tau$, then the leaf is transformed into a decision node. The split test is based on the best attribute.

The basic algorithm can include features like the ability to initialize a VFDT tree with a tree produced by a conventional algorithm, or the ability to deactivate all less promising leaves in the case where the maximum of available memory is reached. Moreover, the memory usage is also minimized, leaving attributes that are less promising.

10.3.1.2 Extensions to VFDT

Since the algorithm of VFDT was published in 2000, several authors have proposed improvements and extensions. The most relevant extend VFDT to deal with continuous attributes and detect and react to drift.

Gama et al. [21] proposed an extension for dealing with continuous attributes. At each leaf, and for each continuous attributes they use a *Btree* to store the attribute values seen so far at the leaf. Each node of the *Btree* contains an attribute value, and the class distribution greater than and less than the attribute value. The authors presented efficient algorithms to compute the information gain of each possible cut-point, traversing the *Btree* once. With the same goal, Jin and Agrawal [26] used a discretization method associated with an interval pruning technique to eliminate less promising intervals, avoiding unnecessary computation of splitting cut-points.

VFDT stores sufficient statistics from hundreds of examples at each leaf node. For the most commonly used splitting functions (like information gain), these statistics assume the form of $P(x_j|y_i)$. Gama et al. [21] proposed using this information to classify test examples. The standard approach to classify a test example is traversing the tree from the root to a leaf and classify the example using the majority class at that leaf. This strategy uses only the class distribution at each leaf $P(y_i)$. Gama et al. [21] classified an example choosing the class that maximizes $P(y_i) \prod P(x_j|y_i)$, that is, using the naive Bayes rule. In [21] the authors presented experimental results that clearly illustrate the advantages of this classification strategy.

Gama et al. [20] presented a variant of VFDT with fast splitting criteria using analytical techniques based on a univariate quadratic discriminant analysis. The sufficient statistics required by the splitting criteria are: the mean and standard variation of each attribute per class. These statistics are fast for incremental update and only require three floats (per attribute/class) to store. The main problem is that the splitting criteria are restricted to two class problems. To solve this restriction, a k classes problem is decomposed into a set of $k * (k - 1)/2$ binary problems generating a forest of binary trees, used collectively to solve the original problem.

VFDT has been extended with the ability to detect and react to time-changing concepts [25]. This is done by learning alternative sub-trees whenever an old one becomes questionable, and replacing the old with the new when the new becomes more accurate.

10.3.1.3 Convergence in the Limit

Domingos and Hulten [13] proved, under realistic assumptions, that the decision trees generated by the Hoeffding Tree Algorithm are asymptotically closed to the one produced by a standard batch algorithm (that is an algorithm that uses all the examples to select the splitting attribute). Domingos and Hulten [13] used the concept of *intentional disagreement* between two decision trees, defined as the probability that one example follow different paths in both trees. They showed that this probability is proportional to δ.

10.3.1.4 Parallel Induction of Decision Trees

VFDT-like algorithms are the state-of-the-art in learning decision trees from data streams. In the case of sensor networks, data are distributed by nature. An open issue is how sensor nodes could collaborate in learning a decision tree. Previous work in parallel induction of decision trees [31,40] can illuminate this problem. In sensor networks, solutions based on vertical data distribution [17] seem the most appropriate. The use of data parallelism in the design of parallel decision tree construction algorithms can be generally described as the execution of the same set of instructions (algorithm) by all processors involved. The parallelism is achieved because each processor is responsible for a distinct set of attributes. Each sensor keeps in its memory only the whole values for the set of attributes assigned to it and the values of the classes. During the evaluation of the possible splits each processor is responsible only for the evaluation of its attributes.

10.3.2 Decision Rules

Ferrer et al. [16] presented FACIL, a classification system based on decision rules that may store up-to-date border examples to avoid unnecessary revisions when virtual drifts are present in data. In the FACIL system, a decision rule r is given by a set of closed intervals which define an hyper-rectangle in the instance space. Rules are stored in different sets according to the associated label. Since no global training window is used but each rule handles a different set of examples (a window per rule), every time a new example arrives the model is updated. In this process, one of three tasks is performed in order:

1. *Positive covering*: x_i is covered by a rule associated with the same label.
2. *Negative covering*: x_i is covered by a rule associated with a different label.
3. *New description*: x_i is not covered by any rule in the model.

Positive covering. First, the rules associated with y_i are visited and the generalization necessary to describe the new example x_i is measured according to the *growth* of the rule. This heuristic gives a rough estimate of the new region of the search space that is taken, biasing in favor of the rule that involves the smallest changes in the minimum number of attributes. While visiting the rules associated with y_i, the one with the minimum growth is marked as *candidate*. However, a rule is taken into account as a possible candidate only if the new example can be seized with a moderate growth (a user parameter).

Negative covering. If x_i is not covered by a rule associated to y_i, then the rest of rules associated with a label $y' \neq y_i$ are visited. If a different label rule r' does not cover x_i, the intersection between r' and the candidate is computed. If such an intersection is not empty, the candidate is rejected. When the first different label rule r'' covering x_i is found, its negative support is increased by one unit, and x_i is added to its window. If the new purity of r'' is smaller than the minimum given by the user, then new consistent rules according to the examples in its window are included in the model. r'' is marked as *unreliable* so that it can not be generalized and has not taken into account to generalize other rules associated with a different label. In addition, its window is reset.

New description. After the above tasks, the candidate rule is generalized if does not intersect with any other rule associated with a label $y' \neq y_i$. If no rule covers the new example and there is not a candidate that can be generalized to cover it, then a maximally specific rule to describe it is generated.

10.3.2.1 Refining and Forgetting Heuristic

FACIL includes a forgetting mechanism that can be either explicit or implicit. Explicit forgetting takes places when the examples are older than an user-defined threshold. Implicit forgetting is performed by removing examples that are no longer relevant as they do not enforce any concept description boundary. When a negative example x in a rule r does not have the same label as the nearest one after the number pe of positive examples that r can store is increased two times since x was covered, the system removes it. Analogously, a positive example is removed if it has not received a different label example as the nearest one after pe is increased by two units.

Finally, to classify a new test example, the system searches the rules that cover it. If there are reliable and unreliable rules covering it, the latter ones are rejected. Consistent rules classify new test examples by covering and inconsistent rules classify them by distance as the nearest neighbor algorithm. If there is no rule covering it, the example is classified based on the label associated with the reliable rule that involves the minimum growth and does not intersect with any different label rule.

10.3.3 Artificial Neural Networks

Artificial neural networks are powerful models that can approximate any continuous function [32] with arbitrary small error with a three-layer network. Craven and Shavlik [10] argued that the inductive bias of neural networks is the most appropriate for sequential and temporal prediction tasks, as most prediction tasks in sensor networks are. The *mauvaise reputation* of neural networks comes from slower learning times and the opaque nature of the model. A decision tree is easier to interpret than a neural network for example.

The process of learning in a neural network involves a gradient-based optimization in the space of the parameters. Iteratively two steps are executed: compute the gradient error function and modify the parameters in the direction of the gradient.

Standard training of neural networks uses several iterations over the training data. The main motivation for this process is the reduced number of examples available with respect to its representation capacity. In the case of sensor networks, the abundance of data eliminates the process of multiple scans of the training data. Each training example is propagated and the error backpropagated through the network only once. The main advantage of this full incremental method used to train the neural network is the ability to process an infinite number of examples that can flow at high speed. Both operations of propagating the example and backpropagating the error through the network are very efficient and can follow high-speed data streams. Another advantage is the smooth adaptation in dynamic data streams where the target function evolves over time.

10.3.4 Support Vector Machines

Sensor networks have a potentially large number of nodes. Each node can send and receive information to its neighbors. This structure is not foreign to the simplified structure of the brain that a neural network (NN) may represent. We can think about each node in a sensor network as a neuron in a NN. This neuron has some threshold for sending information to other neurons; this is similar to a sensor network composed of learning SVMs. Each has to decide what to send and what information or data to request. The SVM is an algorithm based mainly on work performed by Vladimir N. Vapnik and coworkers. It was presented in 1992 and has been the subject of much research since. We look at the algorithm from an application viewpoint and review its characteristics.

The SVM algorithm is a maximal margin algorithm. It seeks to place a hyperplane between classes of points such that the distance between the closest points is maximized. It is equivalent to maximum separation of the distance between the convex hulls enclosing the class member points. Vladimr Vapnik is respected as the researcher who primarily laid the groundwork for the support vector algorithm. The first breakthrough came in 1992 when Boser et al. [3] constructed the SVM learning algorithm as we know it today. The algorithm worked for problems in which the two classes of points were separable by a hyperplane. In the meantime Corinna Cortes was completing her dissertation at AT&T labs, and she and Victor Vapnik worked out the soft margin approach [9], which has a similar structure as a NN. It involves the introduction of slack variables, or error margins that are introduced to absorb errors that are inevitable for non-separable problems. The SVM was primarily constructed to address binary classification problems. This has been adapted by introducing versions of the SVM that can train a multiclassifier concurrently. Other approaches involved the use of voting schemes in which a meta-learner takes the votes from the individual binary classifiers and casts the final vote. A particularly easy voting scheme is the one-against-all voter [37], which for SVMs amounts to training C classifiers and finding the C_i classifier with the hyperplane furthest away from the new point to be tested. For sensor networks, we also have to consider these voting schemes as clusters of nodes can provide information regarding the same phenomena. The SVM has been extended to other learning areas as well. The areas relevant for this work are the extension toward regression and clustering. The regression algorithm extension was pioneered by Vapnik [43] and refined by Smola and Schölkopf [42]. It has been applied in a sensor setting in a thesis by E.M. Jordaan and further discussed by Jordaan and Smits [27]. The regression case is carried out by using the slack variable approach once again. A so-called ε-tube is constructed around the regression line. The width ε constitutes the error free zone, and the points that fall within this zone are regarded as error free. If a point falls outside the tube, then a slack vector approach is introduced, which for the L_2 norm case amounts to minimizing the square distance to the regression line.

Our aim is to provide a framework for addressing inherently distributed inference tasks in sensor networks. This includes, but is not limited to classification, regression, and clustering. Please refer to Chap. 9 for a treatment of clustering. A recent focus is streaming sensor data with or without concept drift. The basic binary classification SVM provides a decision function:

$$f(\boldsymbol{x}, \boldsymbol{\alpha}, b) = \{\pm 1\} = \operatorname{sgn}\left(\sum_{i=1}^{l} \alpha_i y_i k(\boldsymbol{x}_i, \boldsymbol{x}) + b\right) \tag{10.1}$$

for which $\boldsymbol{\alpha}$ has been found by solving this optimization problem:

$$\text{maximize} \quad W(\boldsymbol{\alpha}) = \sum_{i=1}^{l} \alpha_i - \frac{1}{2} \sum_{i=1}^{l} \sum_{j=1}^{l} y_i y_j \alpha_i \alpha_j k(\boldsymbol{x}_i, \boldsymbol{x}_j), \tag{10.2}$$

$$\text{subject to} \quad 0 \le \alpha_i \le C, \quad \text{and} \tag{10.3}$$

$$\sum_{i=1}^{l} y_i \alpha_i = 0. \tag{10.4}$$

The functional output of (10.1) is ± 1, which works as a classification or categorization of the unknown datum $\boldsymbol{x}$ into either the $+$ or $-$ class. An SVM model is constructed by summing a linear combination of training data (historical data) in feature space. Feature space is implicitly constructed by the use of kernels, k. A kernel is a dot product, also called inner product, in a space that is usually not of the same dimensionality as the original input space unless the kernel is just the standard inner product $\langle\, , \rangle$. The constraints (10.3) and (10.4) are the Karush–Kuhn–Tucker (KKT) conditions. The main characteristic of SVM solutions, with respect to the KKT conditions, is for $\alpha_i > 0$: the Lagrange multipliers, $\boldsymbol{\alpha}$, are greater than zero for active constraints in the dual formulated optimization problem. Each of the training data, $\boldsymbol{x}_{i \in 1,\ldots,l}$ is associated with a class label with value $+1$ or -1. This information is contained in the binary variable y_i. The classifying hyperplane induced by (10.1) is offset with a bias parameter b. Estimation of the Lagrange multipliers $\boldsymbol{\alpha}$ is conducted by solving (10.2) subject to the constraints of (10.3) and (10.4). The optimization problem is constructed by enforcing $|f(\boldsymbol{x}, \boldsymbol{\alpha}, b)| = 1$ for the support vectors. Support vectors (SV), are those data, $\boldsymbol{x}_i$, which have active constraints, $\alpha_i > 0$. If the data are not separable by a linear hyperplane in a kernel-induced feature space, then it would not be possible to solve the problem if there was not an upper limit to the values of the active Lagrange multipliers. Consequently, the constraint, C, in (10.3) ensures that the optimization problem in (10.1) remains solvable. In those situations when a Lagrange multiplier is equal to C, the problem is called a soft-margin problem, otherwise it is called a hard-margin problem. The introduction of C is not easy in a sensor network setting. The sensors can be regarded as autonomous. Therefore, it is not particularly easy to work with a parameter which is best optimized using a centralized cross-validation scheme. An important term, which is used extensively is *bound* and *non-bound (NB)*. A training datum, $\boldsymbol{x}_i$ is bound if $\alpha_i = C$. Further introduction to linear classifiers and non-linear extensions such as SVMs can be found in Duda et al. [15].

10.3.4.1 Training and Testing Complexity

Some sensors are as large as a personal PC, but most are small. In the TinyOS community there is even a vision of "smart dust" by Kristofer Pister to have sensors of one cubic millimeter. With this challenge in mind it is important to talk about how a given machine learning algorithm maps to such an environment. An SVM has a computational cost—when trained—that is exactly related to the number of support vectors. Burges introduced the idea of simplified support vectors [4]. Burges and Schölkopf [5] later improved the speed of the SVM with 20-fold gains. The resulting support vectors would generally not be the same as the original support vectors but this was addressed by an elegant and simple solution from Downs et al. [14] in which those support vectors being linearly dependent were eliminated from the solution and the Lagrange multipliers of the remaining support vectors were modified accordingly. It is common to use cross-folding to tune the parameters of the SVM. This further adds cost to the training of the SVM. It has been suggested we use the Lagrange multipliers of a previous SVM to initiate the training of subsequent SVMs [12]. The rationale is that the overall training cost of the SVM is dominated by the cost of the first training on the data set. This is of particular interest when training is performed to reach the leave-one-out error estimate. A straightforward approach [12] is to use the SVM of the Nth cross-fold directly in the case of a zero Lagrange multiplier. If example i to be tested and the Nth cross-fold had a non-zero Lagrange multiplier, then this multiplier value, α_i, is to be redistributed among the rest of the support vectors. One distribution scheme was to redistribute the Lagrange multiplier among the non-bound Lagrange multipliers. The traditional SVM classifies points according to the sign of the functional output. Fung and Mangasarian [18] proposed training a classifier by training two parallel planes pushed as far apart as possible such that the positive and the negative points of each class cluster around the two resulting planes. The algorithm is called a proximal SVM, and it seems to have much lower computational complexity than a traditional SVM, but this has been questioned by Ryan Rifkin in his Ph.D. thesis. The basic SVM was improved by Platt [35] in terms of providing a solution that does not need to cache the full kernel matrix or even parts of it. This basic algorithm was improved by Keerth et al. [28], who used two threshold values instead of one. Several approaches exist to reduce the memory requirements of the SVMs, but Mitra et al. [33] suggested a method that successively draws small samples from a large database while training and keeping an SVM classifier. Similar work has been done to speed up SVM classification by making parallel implementations of the SVM [36]. The approach is to exchange raw data blocks among the nodes in the system and exchange the raw data to obtain the same solutions as the stand-alone SVM algorithm would have done. Chapelle and Vapnik [8] proposed a gradient descent algorithm related directly to the desired generalization performance of the SVM. This allows for finding multiple parameters concurrently. The method is to find the derivatives of the parameters with respect to the margin.

10.4 Distributed Solutions

10.4.1 Kernel Learners in Sensor Networks

The use of statistical learning theory (STL) for designing distributed learning systems will generally involve a combination of appropriate *kernel(s)*, *algorithm(s)*, and *mode(s)*. This natural categorization of design issues is labeled the *KAM* method for designing embedded machine learning systems based on STL. The *KAM* method is introduced in Fig. 10.5 with the elements of each of the three dimensions displayed. First, an appropriate kernel is chosen among the choices the include, but are not limited to: graph, string and tree, set, or vectorial kernels. Second, the algorithm is chosen as one (or more) classification [9], novelty detection [39], regression [38],

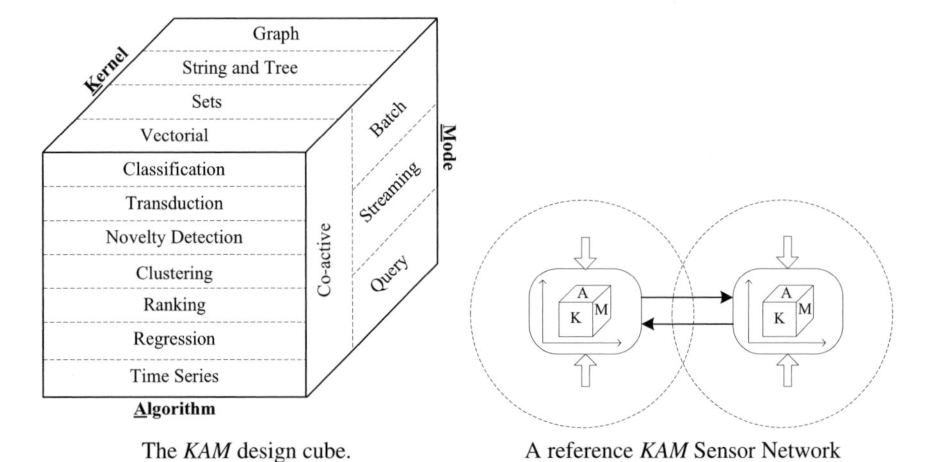

The *KAM* design cube. A reference *KAM* Sensor Network

Fig. 10.5 The *KAM* design cube and a reference *KAM* sensor network

clustering [1,22], ranking [41], or time-series algorithms. Finally, the mode for using the algorithms is chosen as some combination of batch, online, transduction, query, or co-active learning. **The notion of *co-active* learning comes natural in sensor networks.**

10.4.2 SVMs in Sensor Networks

A backpropagation neural network could probably be useful in sensor networks. However, there are some challenges integral in the NN, which include potential overtraining and "loss" of data. While the SVM has a global unique solution that is not subject to overtraining, the capacity of a NN can easily be so big that it can overfit to the data. It is not a new problem in a batch learning scenario (early stopping is one remedy), but in a streaming learning scenario where a model is subject to constant updates it can be a challenge. The sparse solutions produced by SVMs make them important components for distributed data mining problems, especially those related to classification, regression and clustering [34]. In this section we extend the idea of addressing both computational, communication, spatial and temporal constraints. A distributed digraph of SVMs, closely related to the virtual, ν, transductive, and sequential minimal optimization SVMs, is combined with the concept of generalized support vectors to produce two basic types of SVMs: one that filters data to reduce edge exchanges and another that optimizes a cost function using the reduced data set. In addition, the alterations on the support vector set by a controlled relaxation of the Karush–Kuhn–Tucker (KKT) conditions are discussed in the context of employing the recall and precision metrics commonly found in information retrieval settings.

To achieve a level of abstraction from the physical constraints and costs, we propose mapping the categories to a context that is easier to configure in terms of the learning machine. This allows a principled discussion on the parameters of the sensor system and the SVM model that can be modified to address various constraints. The main cost categories are computation, spatial, temporal and communication. Computational costs are mainly perceived as costs associated with CPU cycles for reasons such as limited power supply. Spatial costs are some combination of the cost of using the available memory for the model and how much memory there is available. Temporal costs are those associated with a slow model, and this can translate into a constraint on the maximum time the model can train before it is ready for use.

On resource-constrained devices we often find limited memory availability. Therefore, the data set size of the full training set can be a measure of how well some space constraint is met. Another resource sink could be the lack of kernel caching. The Gram matrix, which holds the kernel outputs $K(\boldsymbol{x}_i, \boldsymbol{x}_j)$, spares some computational units at the expense of a larger memory footprint. To evaluate how well the system filters data, we compare the initial data set size as well as the filtered data.

10.4.3 Data versus Model Exchange Schemes

The distributed inference system exchanges information among the system nodes. There are various ways to execute this exchange. A naive but straightforward way is to replicate all data between all nodes such that each node has the full data set available prior to the training of the model. Alternatively, the inference models can be exchanged among the nodes and allowed to use the data that is available on each node. One aspect that differentiates the SVM from a MLP NN is that the SVM can exchange the data associated with one scalar, which is the Lagrange multiplier α.

SVM exchanges data: It is possible to just exchange the data among the nodes. If all data are exchanged, the problem is the equivalent of having each node equipped with a stand-alone inference problem. It is more relevant to investigate when data are not fully exchanged between the nodes. This is a gradual approach from having one point $\boldsymbol{x}$ on each node N to the case in which all points are represented on all nodes.

Neural networks exchanges models: The other approach is to exchange the models among the nodes and let each model M be updated when it encounters new data on each node. Models can be moved around the nodes systematically or randomly.

It is not completely true that the SVM exchanges only data. In order to save some energy on receiving nodes, it would be an option to tag the exchanged data with the Lagrange multiplier of the sender SVM.

10.5 Emerging Challenges and Future Issues

In sensor networks, data are distributed by nature. In distributed scenarios decisions are collective. Individual nodes access and process local information. In order to achieve a collective decision, they must communicate to neighbor nodes. Instead of exchanging raw data, nodes can send local and partial models and negotiate a common decision with neighbors. From the point of view of knowledge discovery, several challenges and issues emerge:

- The definition of standards and languages for model exchange.[2]
- The incorporation in these languages of the time and space data characteristics.
- Collaboration strategies and conflict resolution.
- Distributed approaches for mining sensitive data become of increase importance: privacy preserving data mining in sensor networks.

Streaming sensor networks pose new challenges. The SVM is one example of a kernel-based learning algorithm which can be applied in a sensor network setting. We hope to have introduced the topic with the intent of establishing this family of learning algorithms for future research. The challenge we face is balancing the

[2] The predictive model markup language is a step in that direction.

resource constraints with the learning requirements of the machine-learning algorithms. As laptops become faster and faster the opposite trend will almost always be evident in sensor networks: we want to achieve real machine learning in an environment constrained by battery life, processor speeds and small memory.

References

[1] A. Ben-Hur, D. Horn, H. Siegelmann, V. Vapnik, Support vector clustering. Journal of Machine Learning Research, 2:125–137, 2001.
[2] A. Bordes, S. Ertekin, J. Weston, L. Bottou, Fast kernel classifiers with online and active learning. Journal of Machine Learning Research, 6:1579–1619, 2005.
[3] B.E. Boser, I. Guyon, V. Vapnik, A training algorithm for optimal margin classifiers. In: Computational Learning Theory, pp. 144–152, 1992.
[4] C.J.C. Burges, Simplified support vector decision rules. In: L. Saitta (Ed.), Proceedings of the 13th International Conference on Machine Learning, pp. 71–77. Morgan Kaufmann, San Mateo, 1996.
[5] C.J.C. Burges, B. Schölkopf, Improving the accuracy and speed of support vector learning machines. In: M. Mozer, M. Jordan, T. Petsche (Eds.), Advances in Neural Information Processing Systems 9, pp. 375–381, MIT Press, Cambridge, 1997.
[6] G. Cauwenberghs, T. Poggio, Incremental and decremental support vector machine learning. In: Proceedings of the 13th Neural Information Processing Systems, pp. 409–415, 2000.
[7] G. Cauwenberghs, T. Poggio, Incremental and decremental support vector machine learning. In: NIPS, pp. 409–415, 2000.
[8] O. Chapelle, V. Vapnik, O. Bousquet, S. Mukherjee, Choosing multiple parameters for support vector machines. Machine Learning, 46(1–3):131–159, 2002.
[9] C. Cortes, V. Vapnik, Support-vector networks. Machine Learning, 20(3):273–297, 1995.
[10] M. Craven, J. Shavlik, Using neural networks for data mining. Future Generation Computer Systems, 13:211–229, 1997.
[11] A.P. Dawid, Statistical theory: the prequential approach. Journal of the Royal Statistical Society A, 147:278–292, 1984.
[12] D. DeCoste, K. Wagstaff, Alpha seeding for support vector machines. In: International Conference on Knowledge Discovery and Data Mining (KDD-2000), 2000.
[13] P. Domingos, G. Hulten, Mining High-Speed Data Streams. In: I. Parsa, R. Ramakrishnan, S. Stolfo (Eds.), Proceedings of the ACM Sixth International Conference on Knowledge Discovery and Data Mining, pp. 71–80. ACM Press, 2000.
[14] T. Downs, K.E. Gates, A. Masters, Exact simplification of support vector solutions. Journal of Machine Learning Research, 2:293–297, 2002.
[15] R.O. Duda, P.E. Hart, D.G. Stork, Pattern Classification, 2nd edn. Wiley–Interscience, New York, 2000.
[16] Ferrer-Troyano, Aguilar-Ruiz, J. Riquelme, Incremental rule learning and border examples selection from numerical data streams. Journal of Universal Computer Science, 11(8):1426–1439, 2005.
[17] A.A. Freitas, S.H. Lavington, Mining Very Large Databases with Parallel Processing. Kluwer Academic, Dordrecht, 1998.
[18] G. Fung, O.L. Mangasarian, Proximal support vector machine classifiers. In: Proceedings of the Seventh ACM SIGKDD International Conference on Knowledge Discovery and Data Mining, pp. 77–86. ACM Press, 2001.
[19] J. Gama, P. Medas, G. Castillo, P. Rodrigues, Learning with drift detection. In: A.L.C. Bazzan, S. Labidi, (Eds.), Advances in Artificial Intelligence—SBIA 2004. Lecture Notes in Computer Science, vol. 3171, pp. 286–295. Springer, Berlin, 2004.

[20] J. Gama, P. Medas, R. Rocha, Forest trees for on-line data. In: Proceedings of the ACM Symposium on Applied Computing, pp. 632–636. ACM Press, 2004.
[21] J. Gama, R. Rocha, P. Medas, Accurate decision trees for mining high-speed data streams. In: Proceedings of the Ninth ACM SIGKDD International Conference on Knowledge Discovery and Data Mining, pp. 523–528. ACM Press, 2003.
[22] M. Girolami, Mercer kernel-based clustering in feature space. IEEE Transactions on Neural Networks, 13(3):780–784, 2001.
[23] J. Gratch, Sequential inductive learning. In: Proceedings of Thirteenth National Conference on Artificial Intelligence, vol. 1, pp. 779–786, 1996.
[24] G. Hulten, P. Domingos, Catching up with the data: research issues in mining data streams. In: Proc. of Workshop on Research Issues in Data Mining and Knowledge Discovery, 2001.
[25] G. Hulten, L. Spencer, P. Domingos, Mining time-changing data streams. In: Proceedings of the 7th ACM SIGKDD International Conference on Knowledge Discovery and Data Mining, pp. 97–106. ACM Press, 2001.
[26] R. Jin, G. Agrawal, Efficient decision tree construction on streaming data. In: P. Domingos, C. Faloutsos (Eds.), Proceedings of the Ninth International Conference on Knowledge Discovery and Data Mining. ACM Press, 2003.
[27] E. Jordaan, G. Smits, Robust outlier detection using SVM regression. In: IEEE International Joint Conference on Neural Networks, vol. 3, pp. 2017–2022, 2004.
[28] S.S. Keerthi, S.K. Shevade, C. Bhattacharyya, K. Murthy, Improvements to Platt's SMO algorithm for SVM classifier design. Neural Computation, 13(3):637–649, 2001.
[29] D. Kifer, S. Ben-David, J. Gehrke, Detecting change in data streams. In: VLDB 04: Proceedings of the 30th International Conference on Very Large Data Bases, pp. 180–191. Morgan Kaufmann, 2004.
[30] R. Klinkenberg, Learning drifting concepts: example selection vs. example weighting. Intelligent Data Analysis, 8(3):281–300, 2004. Special Issue on Incremental Learning Systems Capable of Dealing with Concept Drift.
[31] R. Kufrin, Decision trees on parallel processors. Parallel Processing for Artificial Intelligence, Elsevier, 3:279–306, 1997.
[32] T.M. Mitchell, Machine Learning. McGraw-Hill, New York, 1997.
[33] P. Mitra, C.A. Murthy, S.K. Pal, Data condensation in large databases by incremental learning with support vector machines. In: Proc. Intl. Conf. on Pattern Recognition (ICPR2000), 2000.
[34] R.U. Pedersen, Distributed support vector machine for classification, regression and clustering. In: COLT03 Kernel Impromptu Poster Session, 2003.
[35] J. Platt, Fast Training of Support Vector Machines Using Sequential Minimal Optimization in Advances in Kernel Methods—Support Vector Learning. MIT Press, Cambridge, 1999.
[36] F. Poulet, Multi-way distributed SVM algorithms. In: Parallel and Distributed Computing for Machine Learning. In Conjunction with the 14th European Conference on Machine Learning (ECML'03) and 7th European Conference on Principles and Practice of Knowledge Discovery in Databases (PKDD'03), Cavtat-Dubrovnik, Croatia, 2003.
[37] R. Rifkin, A. Klautau, In defense of one-vs-all classification. Journal of Machine Learning Research, 5:101–141, 2004.
[38] B. Schölkopf, A. Smola, R. Williamson, P. Bartlett, New support vector algorithms. Neural Computation, 12:1083–1121, 2000.
[39] B. Schölkopf, R. Williamson, A. Smola, J. Shawe-Taylor, J. Platt, Support vector method for novelty detection. In: Neural Information Processing Systems 12, 2000.
[40] J.C. Shafer, R. Agrawal, M. Mehta, SPRINT: a scalable parallel classifier for data mining. In: Proc. 22nd Int. Conf. Very Large Databases, pp. 544–555. Morgan Kaufmann, 1996.
[41] J. Shawe-Taylor, N. Christianini, Kernel Methods for Pattern Analysis. Cambridge University Press, Cambridge, 2004.
[42] A. Smola, B. Schölkopf, A tutorial on support vector regression. Neuro COLT2 Technical Report NC2-TR-1998-030, 1998, http://citescer.ist.psu.edu/article/smola98tutorial.html.
[43] V.N. Vapnik, The Nature of Statistical Learning Theory. Springer, Berlin, 1995.

Chapter 11
Tensor Analysis on Multi-aspect Streams

Jimeng Sun, Spiros Papadimitriou, and Philip S. Yu

Abstract Data stream values are often associated with multiple *aspects*. For example, each value from environmental sensors may have an associated type (e.g., temperature, humidity, etc.) as well as location. Aside from time stamp, type and location are the two additional aspects. How to model such streams? How to simultaneously find patterns within and across the multiple aspects? How to do it incrementally in a streaming fashion? In this paper, all these problems are addressed through a general data model, tensor streams, and an effective algorithmic framework, window-based tensor analysis (WTA). Two variations of WTA, independent-window tensor analysis (IW) and moving-window tensor analysis (MW), are presented and evaluated extensively on real data sets. Finally, we illustrate one important application, Multi-Aspect Correlation Analysis (MACA), which uses WTA and we demonstrate its effectiveness on an environmental monitoring application.

11.1 Introduction

Data streams have received attention in different communities due to emerging applications, such as environmental monitoring and sensor networks [20,21]. The data are modeled as a number of co-evolving streams (time series with an increasing length). Most data mining operations need to be re-designed for data streams because of the streaming requirement, i.e., the mining result has to be updated efficiently for the newly arrived data.

J. Sun
Computer Science Department, Carnegie Mellon University, Pittsburgh, PA, USA
e-mail: jimeng@cs.cmu.edu

S. Papadimitriou · P.S. Yu
IBM T.J. Watson Research Center, Hawthorne, NY, USA

S. Papadimitriou
e-mail: spapadim@us.ibm.com

P.S. Yu
e-mail: psyu@us.ibm.com

In the standard stream model, each value is associated with a (time stamp, stream-id) pair. However, the stream-id itself may have some additional structure. For example, it may be decomposed into (location-id, type) $\equiv$ stream-id. We call each such component of the stream-id an *aspect*. The number of discrete values each aspect may take is called its *dimensionality*, e.g., the type aspect has dimensionality 4 and the individual dimensions are temperature, humidity and etc. This additional structure should not be ignored in data exploration tasks, since it may provide additional insights. Thus the typical "flat-world view" may be insufficient. In summary, even though the traditional data stream model is quite general, it cannot easily capture some important aspects of the data, as illustrated through the following motivating example.

Environmental monitoring. Sensitive wildlife and habitats need constant monitoring in a non-intrusive and non-disruptive manner. In such cases, wireless sensors are carefully deployed throughout those habitats to monitor the microclimates such as temperature, humidity and light intensity, as well as the voltage level of the sensors, in real time (see Fig. 11.1). We can view measurements from a particular sensor and from a given location (say temperature around a nesting burrow) as a single stream. In general, there is a large number of such streams that come from different sensor types and locations. Thus, sensing location (which is hard-coded through the hardware sensor id) and sensor type give us two different *aspects* of those streams, which should be clearly differentiated. In particular, it is not sensible to blindly analyze all streams together (for example, it is unusual to compare the humidity at location A to the sensor voltage at location B). Instead, the streams of same type but different locations are often analyzed together for spatial variation patterns. Similarly, the streams at the same location but of different type can be studied for cross-type correlations. In general, a more challenging problem is how to analyze both location and type simultaneously and over time.

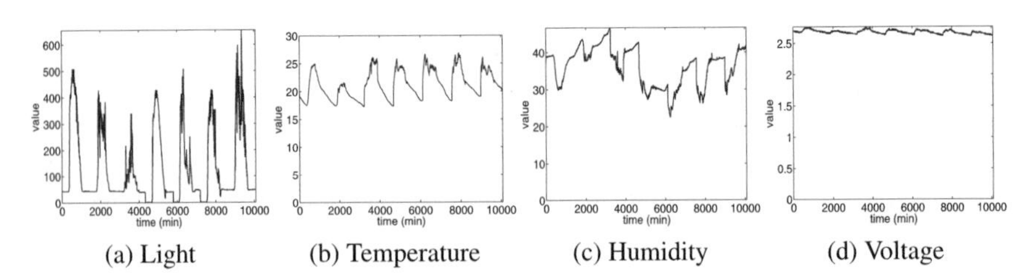

(a) Light (b) Temperature (c) Humidity (d) Voltage

Fig. 11.1 Environmental monitoring: four types over seven days at one location

In this chapter, we tackle the problem at three levels. First, we address the issue of modeling such high-dimensional and multi-aspect streams. More specifically, we present the *tensor stream* model, which generalizes multiple streams into high-order tensors represented using a sequence of multi-arrays. As shown in Fig. 11.6, each matrix on the left is a second-order tensor in the tensor stream. In the environmental monitoring example, each second-order tensor consists of the measurements at a single time stamp from all locations (rows) and types (columns). The tensor representation is closely related to data cube in On-Line Analytical Processing (OLAP). Figure 11.1 lists the corresponding terms.

Table 11.1 Terminology correspondence

This paper	OLAP	Tensor literature
Order	Dimensionality	Order/mode
Aspect/mode	Dimension	Order/mode
Dimension	Attribute value	Dimension

Second, we study how to summarize the tensor stream efficiently. We generalize the moving/sliding window model from a single stream to a tensor stream. Every tensor window includes multiple tensors (an example is highlighted in the tensor stream of Fig. 11.6). Each of these tensors corresponds to the multi-aspect set of measurements associated with one time stamp. Subsequently, using multi-linear analysis [6] which is a generalization of matrix analysis, we propose *window-based tensor analysis (WTA)* for tensor streams, which summarizes the tensor windows efficiently, using small core tensors associated with different projection matrices, where core tensors and projection matrices are analogous to the singular values and singular vectors for a matrix. Two variations of the algorithms for WTA are proposed: (1) *independent-window tensor analysis (IW)* which treats each tensor window independently; and (2) *moving-window tensor analysis (MW)* which exploits the time dependence across neighboring windows to reduce computational cost significantly.

Third, we introduce an important application using WTA, which demonstrates its practical significance. In particular, we describe *Multi-Aspect Correlation Analysis (MACA)*, which *simultaneously* finds correlations within a single aspect and also across multiple aspects.

In the environmental monitoring example, Fig. 11.2 shows two factors across three aspects (modes): time, location and type. The technical formulation of those factors is discussed in Sect. 11.6. Intuitively, each factor corresponds to a trend (either normal or abnormal) and consists of a global "importance" weight (or scaling constant) of this trend, a pattern across time summarizing the "typical behavior" for this trend and, finally, one set of weights for each aspect[1] (representing their participation in this trend).

Figures 11.2(a1, b1, c1) show the three components (one for each aspect) of the first factor, which is the main trend. If we read the components independently, Fig. 11.2(a1) shows the daily periodic pattern over time (high activation during the day, low activation at night); (b1) shows the participation weights of all sensor locations, where the weights seem to be uniformly spread out; (c1) shows that light, temperature and voltage are positively correlated (all positive values) while they are

[1] Time is also an aspect. However, the time aspect and its associated "weights" have a special significance and interpretation.

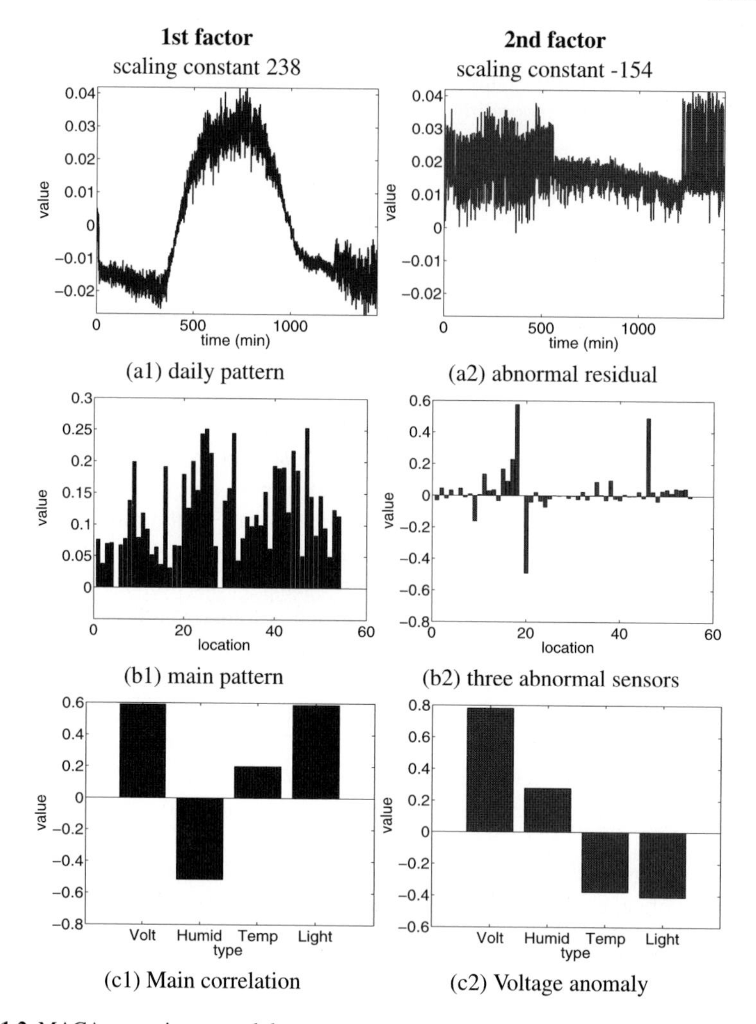

Fig. 11.2 MACA on environmental data

anti-correlated with humidity (negative value). All of those patterns can be confirmed from the raw data (partly shown in Fig. 11.1). All three components act together scaled by the constant factor 238 to form an approximation of the data.

Part of the residual of that approximation is shown as the second factor in Figs. 11.2(a2, b2, c2), which corresponds to some major anomalies. More specifically, Fig. 11.2(a2) suggests that this pattern is uniform across all time, without a clear trend as in (a1); (b2) shows that two sensors have strong positive weights while one other sensor has a dominating low weight; (c2) tells us that this happened mainly at voltage which has the most prominent weight. The residual (i.e., the information remaining after the first factor is subtracted) is in turn approximated by combining all three components and scaling by constant -154. In layman terms,

Table 11.2 Description of notation

Symbol	Description
$\mathbf{v}$	A vector (lowercase bold)
$\mathbf{v}(i)$	The i-element of vector $\mathbf{v}$
$\mathbf{U}$	A matrix (uppercase bold)
$\mathbf{U}^T$	The transpose of $\mathbf{A}$
$\mathbf{U}_i \mid_{i=1}^{n}$	A sequence of N matrices $\mathbf{A}_1, \ldots, \mathbf{A}_n$
$\mathbf{U}(i, j)$	The entry (i, j) of $\mathbf{A}$
$\mathbf{U}(i, :)$ or $\mathbf{U}(:, i)$	ith row or column of $\mathbf{A}$
$\mathcal{D}$	A tensor (calligraphic style)
$\mathcal{D}(i_1, \ldots, i_M)$	The element of $\mathcal{X}$ with index $(i_1, \ldots, i_M)$
M	The order of the tensor
W	The window size of the tensor window
N_i	The dimensionality of the ith mode of input tensors
R_i	The dimensionality of the ith mode of core tensors
WTA	Window-based tensor analysis
IW	Independent-window tensor analysis
MW	Moving-window tensor analysis
MACA	Multi-Aspect Correlation Analysis

two sensors have abnormally low voltage compared to the rest; while one other sensor has unusually high voltage. Again all three cases are confirmed from the data.

In summary, our main contributions are the following:

- *Data model*: We introduce tensor streams to deal with high-dimensional and multi-aspect streams.
- *Algorithmic framework*: We propose window-based tensor analysis (WTA) to effectively extract core patterns from tensor streams.
- *Application*: Based on WTA, multi-aspect correlation analysis (MACA) is presented to simultaneously compute the correlation within and across all aspects.

The rest of the paper is organized as follows: Sect. 11.2 briefly describes the background. Section 11.3 presents some key definitions and problems. Section 11.4 describes our proposed methods. Sections 11.5 and 11.6 present the experimental evaluation and case study, respectively. Finally, Sect. 11.7 describes some other related work and Sect. 11.8 concludes.

11.2 Background

In the following, we use lowercase bold letters for column vectors $\mathbf{x}$, uppercase bold for matrices $\mathbf{U}$, and calligraphic uppercase for tensor $\mathcal{D}$. The ith column and row of matrix $\mathbf{U}$ is denoted as $\mathbf{U}(:, i)$ and $\mathbf{U}(i, :)$, respectively.

11.2.1 Singular Value Decomposition (SVD)

SVD is a matrix operation that decomposes a matrix $\mathbf{X}$ into a product of three matrices. Formally, we have $\mathbf{X} = \mathbf{U}\mathbf{\Sigma}\mathbf{V}^T$ where $\mathbf{X} \in \mathbb{R}^{m \times n}$ is a rank r matrix, $\mathbf{U} \in \mathbb{R}^{m \times r}$ and $\mathbf{V} \in \mathbb{R}^{n \times r}$ consists left and right singular vectors and $\mathbf{\Sigma}$ is a diagonal matrix with singular values on the diagonal. As shown in Figure 11.3, SVD reveals the structure of the matrix by basis transformation on both the column and row space.

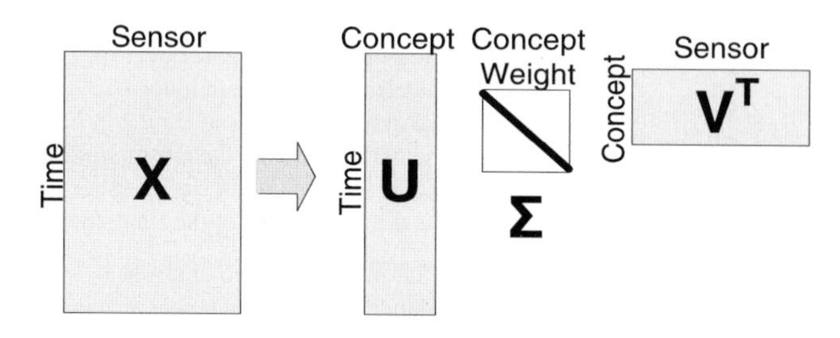

Fig. 11.3 SVD decomposes "time vs. sensor-id" matrix into three matrices: $\mathbf{U}$ is the "time to concept" matrix that gives soft-clusters of time stamps; similarly, $\mathbf{V}$ gives the clusters on sensors; the diagonal elements in $\mathbf{\Sigma}$ provide the weights for different concepts

Two important applications follow from SVD: (1) *Principal component analysis (PCA)*: Given a matrix $\mathbf{X} \in \mathbb{R}^{m \times n}$ where every row is a n-dimensional point, PCA projects all m points into r dimensional subspace. Computationally (assuming the rows are zero mean), PCA performs SVD (i.e., $\mathbf{X} = \mathbf{U}\mathbf{\Sigma}\mathbf{V}^T$), then groups $\mathbf{Y} = \mathbf{U}\mathbf{\Sigma}$ as the r-dimensional points and $\mathbf{V}$ specifies the subspace. (2) *Latent Semantic Indexing (LSI)*: The matrices $\mathbf{U}$ and $\mathbf{V}$ from SVD can be used to simultaneously cluster the rows and columns of the matrix. More specifically, the cluster is formed by picking the large entries (in absolute value) of each column of $\mathbf{U}$ and $\mathbf{V}$. For example in Fig. 11.3, assuming the first columns of $\mathbf{U}$ and $\mathbf{V}$, denoted as $\mathbf{U}(:, 1)$ and $\mathbf{V}(:, 1)$, are $[.707, .707, 0, 0]^T$ and $[.707, 0, .707]^T$ respectively, then the time stamp 1 and 2 form a time cluster, similarly sensor 1 and 3 forms the corresponding sensor cluster. Moreover, this time cluster (time stamp 1, 2) is associated with the sensor cluster (sensor 1, 3).

11.2.2 Multi-Linear Analysis

As mentioned, a tensor of order M closely resembles a data cube with M dimensions. Formally, we write an Mth order tensor $\mathcal{X} \in \mathbb{R}^{N_1 \times \cdots \times N_M}$ as $\mathcal{X}_{[N_1, \ldots, N_M]}$, where N_i $(1 \le i \le M)$ is the *dimensionality* of the ith mode ("dimension" in OLAP terminology). For brevity, we often omit the subscript $[N_1, \ldots, N_M]$.

Definition 1 (Matricizing or Unfolding). The mode-d matricizing or matrix unfolding of an Mth order tensor $\mathcal{X} \in \mathbb{R}^{N_1 \times \cdots \times N_M}$ are vectors in $\mathbb{R}^{N_d}$ obtained by keeping index d fixed and varying the other indices. Therefore, the mode-d matricizing $\mathbf{X}_{(d)}$ is in $\mathbb{R}^{(\prod_{i \neq d} N_i) \times N_d}$.

The mode-d matricizing $\mathcal{X}$ is denoted as unfold$(\mathcal{X}, d) = \mathbf{X}_{(d)}$. Similarly, the inverse operation is denoted as fold$(\mathbf{X}_{(d)})$. In particular, we have $\mathcal{X} = \text{fold}(\text{unfold}(\mathcal{X}, d))$. Figure 11.4 shows an example of mode-1 matricizing of a third order tensor $\mathcal{X} \in \mathbb{R}^{N_1 \times N_2 \times N_3}$ to the $(N_2 \times N_3) \times N_1$-matrix $\mathbf{X}_{(1)}$. Note that the shaded area of $\mathbf{X}_{(1)}$ in Fig. 11.4 is the slice of the third mode along the second dimension.

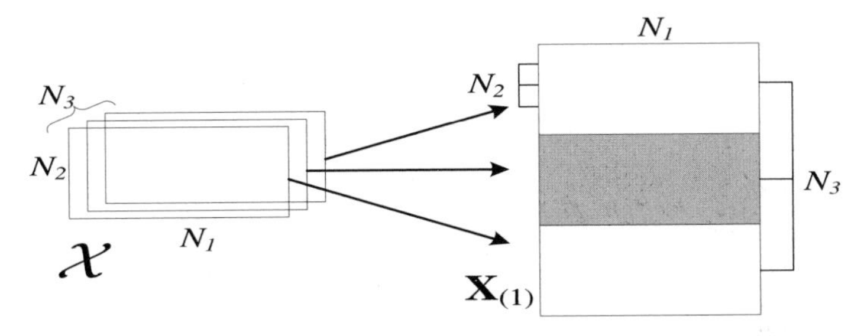

Fig. 11.4 Third order tensor $\mathcal{X} \in \mathbb{R}^{N_1 \times N_2 \times N_3}$ is matricized along the first mode, into a matrix $\mathbf{X}_{(1)} \in \mathbb{R}^{(N_2 \times N_3) \times N_1}$. The shaded area is the slice of the third mode along the second dimension

Definition 2 (Mode Product). The mode product $\mathcal{X} \times_d \mathbf{U}$ of a tensor $\mathcal{X} \in \mathbb{R}^{N_1 \times \cdots \times N_M}$ and a matrix $\mathbf{U} \in \mathbb{R}^{N_i \times N'}$ is the tensor in $\mathbb{R}^{N_1 \times \cdots \times N_{i-1} \times N' \times N_{i+1} \times \cdots \times N_M}$ defined by:

$$
\begin{aligned}
&(\mathcal{X} \times_d \mathbf{U})(i_1, \ldots, i_{d-1}, j, i_{d+1}, \ldots, i_M) \\
&= \sum_{i_d=1}^{N_i} \mathcal{X}(i_1, \ldots, i_{d-1}, i_d, i_{d+1}, \ldots, i_M)\mathbf{U}(i_d, j)
\end{aligned}
\tag{11.1}
$$

for all index values.

Figure 11.5 shows an example of third order tensor $\mathcal{X}$ mode-1 multiplies a matrix $\mathbf{U}$. The process is equivalent to first matricizing $\mathcal{X}$ along mode-1 as $\mathbf{X}_1$, then doing matrix multiplication of $\mathbf{X}_1$ and $\mathbf{U}$, finally folding the result back as a tensor.

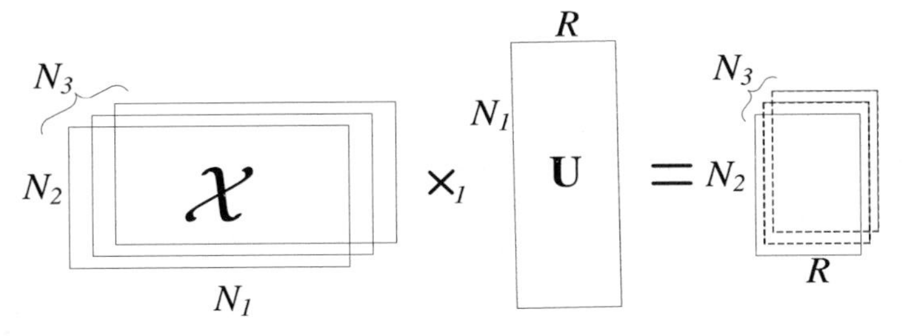

Fig. 11.5 Third order tensor $\mathcal{X}_{[N_1, N_2, N_3]} \times_1 \mathbf{U}$ results in a new tensor in $\mathbb{R}^{R \times N_2 \times N_3}$

In general, a tensor $\mathcal{X} \in \mathbb{R}^{N_1 \times \cdots \times N_M}$ can multiply a sequence of matrices $\mathbf{U}_i|_{i=1}^{M} \in \mathbb{R}^{N_i \times R_i}$ as: $\mathcal{X} \times_1 \mathbf{U}_1 \cdots \times_M \mathbf{U}_M \in \mathbb{R}^{R_1 \times \cdots \times R_M}$, which can be written as $\mathcal{X} \prod_{i=1}^{M} \times_i \mathbf{U}_i$ for clarity. Furthermore, the notation for $\mathcal{X} \times_1 \mathbf{U}_1 \cdots \times_{i-1} \mathbf{U}_{i-1} \times_{i+1} \mathbf{U}_{i+1} \cdots \times_M \mathbf{U}_M$ (i.e. multiplication of all $\mathbf{U}_j$s except the ith) is simplified as $\mathcal{X} \prod_{j \neq i} \times_j \mathbf{U}_j$.

Definition 3 (Tensor Product). The tensor product $\mathcal{X} \otimes \mathcal{Y}$ of a tensor $\mathcal{X} \in \mathbb{R}^{N_1 \times \cdots \times N_M}$ and another tensor $\mathcal{Y} \in \mathbb{R}^{P_1 \times \cdots \times P_Q}$ is the tensor in $\mathbb{R}^{N_1 \times \cdots \times N_M \times P_1 \times \cdots \times P_Q}$ defined by:

$$(\mathcal{X} \otimes \mathcal{Y})(i_1, \ldots, i_M, j_1, \ldots, j_Q) = \mathcal{X}(i_1, \ldots, i_M) \cdot \mathcal{Y}(j_1, \ldots, j_Q) \tag{11.2}$$

for all indices.

Intuitively, the tensor product $\mathcal{X} \otimes \mathcal{Y}$ creates a copy of tensor $\mathcal{Y}$ for each element of tensor $\mathcal{X}$ then multiplies each copy of $\mathcal{Y}$ with the corresponding element in $\mathcal{X}$, which results in a higher order tensor. In fact, the tensor product of two vectors **x** and **y** is the outer product where results in a rank-1 matrix (i.e., a second-order tensor).

Definition 4 (Rank-($R_1, \ldots, R_M$) approximation). Given a tensor $\mathcal{X} \in \mathbb{R}^{N_1 \times \cdots \times N_M}$, a tensor $\tilde{\mathcal{X}} \in \mathbb{R}^{N_1 \times \cdots \times N_M}$ with $\text{rank}(\tilde{\mathbf{X}}_{(d)}) = R_d$ for $1 \leq d \leq M$, that minimizes the least-squares cost $\|\mathcal{X} - \tilde{\mathcal{X}}\|_F^2$, is the best rank-$(R_1, \ldots, R_M)$ approximation of $\mathcal{X}$.[2]

The best rank approximation $\tilde{\mathcal{X}}$ is $\mathcal{X} \prod_{i=1}^{M} \times_i \mathbf{U}_i \mathbf{U}_i^T$, where $\mathbf{U}_l|_{l=1}^{M} \in \mathbb{R}^{N_l \times R_l}$ are the projection matrices.

11.3 Problem Formulation

In this section, we first formally define the notions of *tensor stream* and *tensor window*. Then we formulate two versions of window-based tensor analysis (WTA). Recall that a tensor mode corresponds to an aspect of the data streams.

Definition 5 (Tensor stream). A sequence of Mth order tensors $\mathcal{D}_1, \ldots, \mathcal{D}_n$, where each $\mathcal{D}_i \in \mathbb{R}^{N_1 \times \cdots \times N_M}$ $(1 \leq i \leq n)$ and n is an integer that increases with time, is called a tensor stream denoted as $\{\mathcal{D}_i \mid 1 \leq i \leq n, n \to \infty\}$.

We can consider a tensor stream is built incrementally over time. The most recent tensor in the stream is $\mathcal{D}_n$. In the environmental monitoring example, a new second-order tensor (as shown in Fig. 11.6) arrives every minute.

Definition 6 (Tensor window). A tensor window $\mathcal{D}(n, W)$ consists of a subset of a tensor stream ending at time n with size W. Formally $\mathcal{D}(n, w) \in \mathbb{R}^{W \times N_1 \times \cdots \times N_M} = \{\mathcal{D}_{n-w+1}, \ldots, \mathcal{D}_n\}$ where each $\mathcal{D}_i \in \mathbb{R}^{N_1 \times \cdots \times N_M}$.

[2] The square Frobenius norm is defined as $\|\mathcal{X}\|_F^2 = \sum_{i=1}^{N_1} \cdots \sum_{i=1}^{N_M} \mathcal{X}(i_1, \ldots, i_M)^2$.

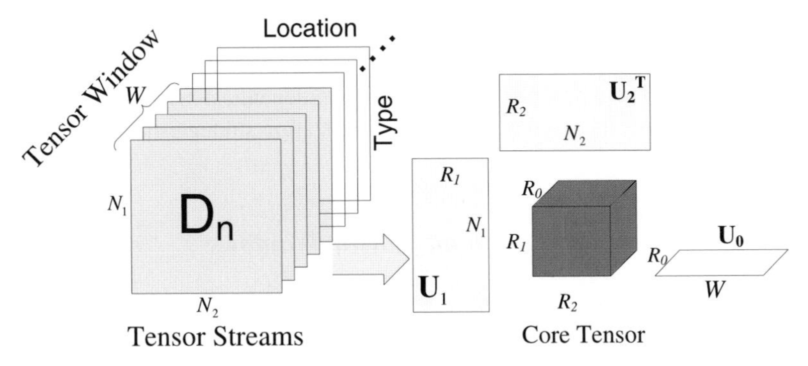

Fig. 11.6 Window-based tensor analysis

A tensor window localizes the tensor stream into a (smaller) tensor sequence with cardinality W and at particular time n. The current tensor window refers to the window ending at the current time. Notice that the tensor window is a natural high-order generalization of the sliding window model in data streams.

The goal of *window-based tensor analysis (WTA)* is to incrementally extract patterns from high-dimensional and multi-aspect streams. In this chapter, we formulate this pattern-extraction process as a dimensionality reduction problem on tensor windows. Two versions of WTA are presented as follows:

Problem 1 (Independent-window tensor analysis (IW)). Given a tensor window $\mathcal{D} \in \mathbb{R}^{W \times N_1 \times \cdots \times N_M}$, find the projection matrices $\mathbf{U}_0 \in \mathbb{R}^{W \times R_0}$ and $\mathbf{U}_i \in \mathbb{R}^{N_i \times R_i} \,|_{i=1}^{M}$ such that the reconstruction error is minimized: $e = \|\mathcal{D} - \mathcal{D}\prod_{i=0}^{M} \times_i (\mathbf{U}_i\mathbf{U}_i^T)\|_F^2$.

The core tensor $\mathcal{Y}$ is defined as $\mathcal{D}\prod_{i=0}^{M} \times_i \mathbf{U}_i$. Intuitively, a projection matrix specifies the "concepts" along a given aspect/mode, while the core tensor consists of the "concept" association across all aspects/modes. Note that for second-order tensors (i.e., matrices), the core tensor is the diagonal matrix Σ of singular values.

Problem 2 (Moving-window tensor analysis (MW)). Given the current tensor window $\mathcal{D}(n, W) \in \mathbb{R}^{W \times N_1 \times \cdots \times N_M}$ and the old result for $\mathcal{D}(n-1, W)$, find the new projection matrices $\mathbf{U}_0 \in \mathbb{R}^{W \times R_{0,n}}$ and $U_i \in \mathbb{R}^{N_i \times R_{i,n}} \,|_{i=1}^{M}$ such that the reconstruction error is minimized: $e = \|\mathcal{D}(n, W) - \mathcal{D}(n, W)\prod_{i=0}^{M} \times_i (\mathbf{U}_i\mathbf{U}_i^T)\|_F^2$

Finally, we illustrate a mining application using the output of WTA in Sect. 11.6.

11.4 Window-Based Tensor Analysis

The tensor stream, like other data streams, can be highly non-linear and unstable over time. One common way to deal with such concept drift is to partition the entire stream into time windows and then fit local models for individual windows.

To do so, Sect. 11.4.1 first introduces the goal and insight of the general window-based tensor analysis, where we point out the importance of good initialization for the algorithm. Next we propose two algorithms, independent-window and moving-window tensor analysis based on different initialization strategies in Sect. 11.4.2.

11.4.1 Iterative Optimization on Tensor Windows

This section first presents the general procedure for tensor analysis. Then we illustrate the challenges for streams and suggest the direction towards which the algorithms have to be developed.

The goal of tensor analysis is to find the set of orthonormal projection matrices $\mathbf{U}_i|_{i=0}^{M}$ that minimize the reconstruction error $d(\mathcal{D}, \tilde{\mathcal{D}})$, where $d(\cdot,\cdot)$ is a divergence function between two tensors; $\mathcal{D}$ is the input tensor; $\tilde{\mathcal{D}}$ is the approximation tensor defined as $\mathcal{D}\prod_{i=0}^{M} \times_i (\mathbf{U}_i\mathbf{U}_i^T)$.

These types of problems can usually be solved through iterative optimization techniques, such as the power method [10], alternating least squares [26] and expectation maximization (EM) [7]. The principle is to optimize parameters one at a time by fixing all the rest. The benefit comes from the simplicity and robustness of the algorithms. As an additional side-benefit, missing values can be filled in during the iterations, at little additional cost (similar to EM-based imputation [17]) . An iterative meta-algorithm for window-based tensor analysis is shown in Fig. 11.7.

Input:
The tensor window $\mathcal{D} \in \mathbb{R}^{W \times N_1 \times \cdots \times N_M}$
The dimensionality of the output tensors $\mathcal{Y} \in \mathbb{R}^{R_0 \times \cdots \times R_M}$.
Output:
The projection matrix $\mathbf{U}_0 \in \mathbb{R}^{W \times R_0}$, $\mathbf{U}_i|_{i=1}^{M} \in \mathbb{R}^{N_i \times R_i}$
and the core tensor $\mathcal{Y}$.
Algorithm:
1. Initialize $\mathbf{U}_i\ |_{i=0}^{M}$
2. Conduct 3–5 iteratively
3. For $k = 0$ to M
4. Fix $\mathbf{U}_i$ for $i \neq k$ and find the $\mathbf{U}_k$ that minimizes $d(\mathcal{D}, \mathcal{D}\prod_{i=0}^{M} \times_i (\mathbf{U}_i\mathbf{U}_i^T))$
5. Check convergence
6. Calculate the core tensor $\mathcal{Y} = \mathcal{D}\prod_{i=0}^{M} \times_i \mathbf{U}_i$

Fig. 11.7 Iterative tensor analysis

To instantiate this meta-algorithm, three things have to be specified:
Initialization condition: This turns out to be the crucial component for data streams. Different schemes for this are presented in Sect. 11.4.2.

Optimization strategy: This is closely related to the divergence function $d(\cdot, \cdot)$. Gradient descent type of methods can be developed in most of cases. However, in this chapter, we use the square Frobenius norm $\|\cdot\|_F^2$ as the divergence function, which naturally leads to a simpler and faster iterated method, alternating least squares. More specifically, line 4 of Fig. 11.7 is replaced by the following steps:

1. Construct $\mathcal{D}^d = \mathcal{D}(\prod_{i \neq d} \times_i \mathbf{U}_i) \in \mathbb{R}^{R_0 \times \cdots \times N_d \times \cdots \times R_M}$
2. unfold$(\mathcal{D}^d, d)$ as $\mathbf{D}_{(d)} \in \mathbb{R}^{N_d \times (\prod_{k \neq d} R_k)}$
3. Construct variance matrix $\mathbf{C}_d = \mathbf{D}_{(d)}^T \mathbf{D}_{(d)}$
4. Compute $\mathbf{U}_d$ by diagonalizing $\mathbf{C}_d$

Convergence checking: We use the standard approach of monitoring the change of the projection matrices until it is sufficiently small. Formally, the change on ith mode is quantified by trace$(||\mathbf{U}_{i,\text{new}}^T \mathbf{U}_{i,\text{old}}| - \mathbf{I}|)$.

11.4.2 Independent and Moving Window Tensor Analysis

In a streaming setting, WTA requires quick updates when a new tensor window arrives. Ultimately, this reduces to quickly setting a good initial condition for the iterative algorithm. In this section, we first introduce independent-window tensor analysis (IW) as the baseline algorithm. Then the moving-window tensor analysis (MW) is presented that exploits the time dependence structure to quickly set a good initial condition, thereby significantly reducing the computational cost. Finally, we provide some practical guidance on choosing the size of the core tensors.

In general, any scheme has to balance the quality of the initial condition with the cost of obtaining it. A simple initialization can be fast, but it may require a large number of iterations until convergence. On the other hand, a complex scheme can give a good initial condition leading to much fewer iterations, but the overall benefit can still be diminished due to the time-consuming initialization.

11.4.2.1 Independent Window Tensor Analysis (IW)

IW is a simple way to deal with tensor windows by fitting the model independently. At every time stamp a tensor window $\mathcal{D}(n, W)$ is formed, which includes the current tensor $\mathcal{D}_n$ and $W - 1$ old tensors. Then we can apply the alternating least squares method (Fig. 11.7) on $\mathcal{D}(n, W)$ to compute the projection matrices $\mathbf{U}_i|_{i=0}^M$. The idea is illustrated in Fig. 11.8. The projection matrices $\mathbf{U}_i$ can, in theory, be any orthonormal matrices. For instance, we initialize $\mathbf{U}_i$ to be a $N_i \times R_i$ truncated identity matrix in the experiment, which leads to extremely fast initialization of the projection matrices. However, the number of iterations until convergence can be large.

An initialization which may at first seem better can be easily achieved as follows: (1) unfold tensor along each mode, i.e., $\mathbf{X}_{(d)} = $ unfold$(\mathcal{X}, d)$; (2) compute the variance matrix of that mode $\mathbf{C}_d = \mathbf{X}_{(d)}^T \mathbf{X}_{(d)} \in \mathbb{R}^{N_d \times N_d}$; (3) find the matrix of the top

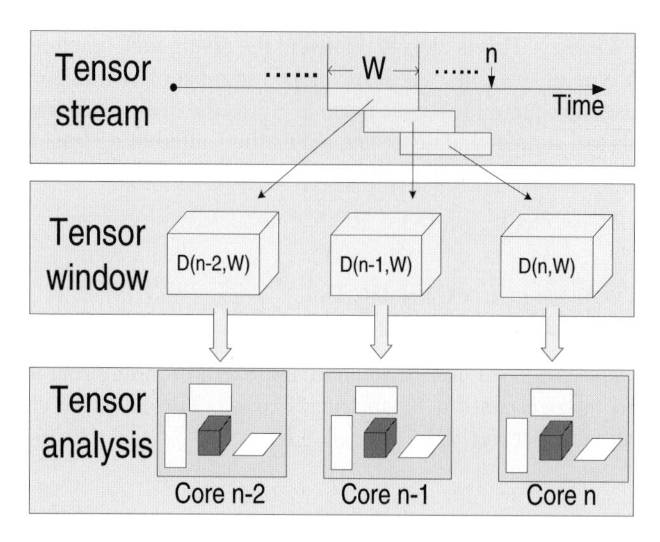

Fig. 11.8 IW computes the core tensors and projection matrices for every tensor window separately, despite that fact there can be overlaps between tensor windows

R_i eigenvectors of $\mathbf{C}_d$ and use that as the initial value of $\mathbf{U}_i$. However, in a streaming environment, the computational cost of this initialization (mainly from computing $\mathbf{C}_d$) can easily offset the benefit of a smaller number of successive iterations. Next we present an incremental initialization scheme that can obtain good starting points with low overhead.

11.4.2.2 Moving-window tensor analysis (MW)

MW explicitly uses the overlapping information of the two consecutive tensor windows to update the variance matrices $\mathbf{C}_d|_{d=1}^{M}$. More specifically, given a tensor window $\mathcal{D}(n, W) \in \mathbb{R}^{W \times N_1 \times \cdots \times N_M}$, we have $M+1$ variance matrices $\mathbf{C}_i|_{i=0}^{M}$, one for each mode. Note that the current window $\mathcal{D}(n, W)$ removes an old tensor $\mathcal{D}_{n-W}$ and includes a new tensor $\mathcal{D}_n$, compared to the previous window $\mathcal{D}(n-1, W)$ (see Fig. 11.9).

Update modes 1 to M: For all but the time mode, the variance matrix is as follows:

$$\mathbf{C}_d^{\text{old}} = \begin{bmatrix} \mathbf{X} \\ \mathbf{D} \end{bmatrix}^T \begin{bmatrix} \mathbf{X} \\ \mathbf{D} \end{bmatrix} = \mathbf{X}^T\mathbf{X} + \mathbf{D}^T\mathbf{D},$$

where $\mathbf{X}$ is the unfolding matrix of the old tensor $\mathcal{D}_{n-W}$ and $\mathbf{D}$ is the unfolding matrix of tensor window $\mathcal{D}(n-1, W-1)$ (i.e., the overlapping part of the two consecutive tensor windows). Similarly, $\mathbf{C}_d^{\text{new}} = \mathbf{D}^T\mathbf{D} + \mathbf{Y}^T\mathbf{Y}$, where $\mathbf{Y}$ is the unfolding matrix of the new tensor $\mathcal{D}_n$. As a result, the update can be easily achieved as follows:

$$\mathbf{C}_d \leftarrow \mathbf{C}_d - \mathbf{D}_{n-W,(d)}^T \mathbf{D}_{n-W,(d)} + \mathbf{D}_{n,(d)}^T \mathbf{D}_{n,(d)},$$

where $\mathbf{D}_{n-W,(d)}(\mathbf{D}_{n,(d)})$ is the mode-d unfolding matrix of tensor $\mathcal{D}_{n-W}(\mathcal{D}_n)$—see Fig. 11.9. Intuitively, the variance matrix can be updated easily when adding or deleting rows from an unfolded matrix, since all computation only involves the added and deleted rows.

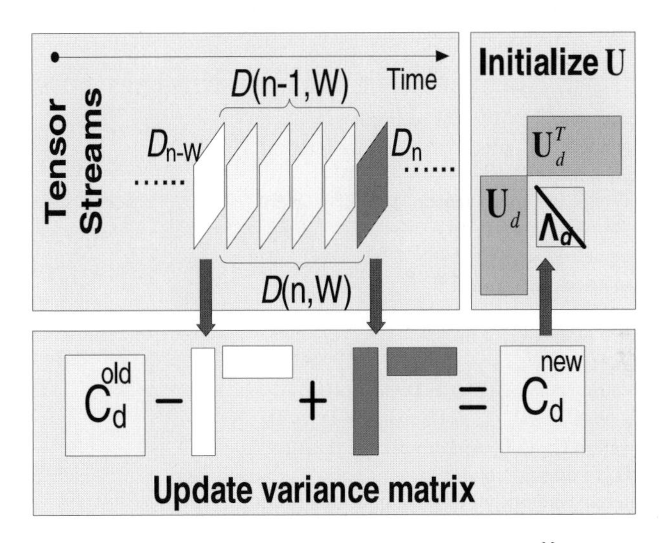

Fig. 11.9 The key of MW is to initialize the projection matrices $\mathbf{U}_d|_{d=1}^M$ by diagonalizing the variance matrices $\mathbf{C}_d$, which are incrementally maintained. Note that $\mathbf{U}_0$ for the time mode is not initialized, since it is different from the other modes

Update mode 0: For the time mode, every update involves adding and removing a column instead of rows. More specifically, unfolding tensor window $\mathcal{D}_{n-W}, \ldots, \mathcal{D}_n$ along the time mode gives a matrix $[\mathbf{x}\,\mathbf{D}\,\mathbf{y}]$ where $\mathbf{x}$ and $\mathbf{y}$ are the vectorizations of the old tensor $\mathcal{D}_{n-W}$ and the new tensor $\mathcal{D}_n$, respectively; $\mathbf{D}$ is again the overlapping part of two tensor windows. The old variance matrix for tensor window $\mathcal{D}(n-1, W)$ is the following:

$$\mathbf{C}_0^{\text{old}} = [\mathbf{x}\,\mathbf{D}]^T[\mathbf{x}\,\mathbf{D}] = \begin{bmatrix} \mathbf{x}^T\mathbf{x} & \mathbf{x}^T\mathbf{D} \\ \mathbf{D}^T\mathbf{x} & \mathbf{D}^T\mathbf{D} \end{bmatrix}.$$

Similarly, the new variance matrix for tensor window $\mathcal{D}(n, W)$ can be incrementally computed as follows:

$$\mathbf{C}_0^{\text{new}} = \begin{bmatrix} \mathbf{D}^T\mathbf{D} & \mathbf{D}^T\mathbf{y} \\ \mathbf{y}^T\mathbf{D} & \mathbf{y}^T\mathbf{y} \end{bmatrix},$$

where $\mathbf{D}^T\mathbf{D}$ is the subset of $\mathbf{C}_0^{\text{old}}$ which can be obtained without any computation; $\mathbf{y}^T\mathbf{y}$ is the inner product of the new data. Unfortunately, $\mathbf{D}^T\mathbf{y}$ involves both the new tensor and all the other tensors in the window. Therefore, the update to variance matrix is not local to the added and deleted tensor, but has a global effect over the entire window.

Fortunately, the iterative algorithm actually only needs initialization for all but one mode in order to start. Therefore, after initialization of the other modes, the iterated update starts from the time mode and proceeds until convergence. This gives both quick convergence and fast initialization. The pseudo-code is listed in Fig. 11.10.

Input:
The new tensor $\mathcal{D}_n \in \mathbb{R}^{N_1 \times \cdots \times N_M}$ for inclusion
The old tensor window $\mathcal{D}_{n-W} \in \mathbb{R}^{N_1 \times \cdots \times N_M}$ for removal
The old variance matrices $\mathbf{C}_d|_{d=1}^{M}$
The dimensionality of the output tensors $\mathcal{Y} \in \mathbb{R}^{R_0 \times \cdots \times R_M}$
Output:
The new variance matrices $\mathbf{C}_d|_{d=1}^{M}$
The projection matrices $\mathbf{U}_i|_{i=0}^{M} \in \mathbb{R}^{N_i \times R_i}$
Algorithm:
// Initialize every mode except time
1. For $d = 1$ to M
2. Mode-d matricize $\mathcal{D}_{n-W}(\mathcal{D}_n)$ as $\mathbf{D}_{n-W,(d)}(\mathbf{D}_{n,(d)})$
4. Update $\mathbf{C}_d \leftarrow \mathbf{C}_d - \mathbf{D}_{n-W,(d)}^T \mathbf{D}_{n-W,(d)} + \mathbf{D}_{n,(d)}^T \mathbf{D}_{n,(d)}$
5. Diagonalization $\mathbf{C}_d = \mathbf{U}_d \mathbf{\Lambda}_d \mathbf{U}_d^T$
6. Truncate $\mathbf{U}_d$ to first R_d columns
7. Apply the iterative algorithm with the new initialization

Fig. 11.10 Moving-window tensor analysis (MW)

Batch update: Often the updates for window-based tensor analysis consist of more than one tensors. Either the input tensors are coming in batches, or the processing unit waits until enough new tensors appear and then triggers the updates. In terms of the algorithm, the only difference is that the update to variance matrices involves more than two tensors (line 4 of Fig. 11.10).

11.4.2.3 Choosing R

The sizes of the core tensor $R_0 \times \cdots \times R_M$ are system parameters. In practice, there are usually two mechanisms for setting them:

- *Reconstruction-guided*: The larger R_i's are, the better approximation we can get. But the computational and storage cost will increase accordingly (see Fig. 11.12 for detailed discussion). Therefore, users can set a desirable threshold for the error, then the proper sizes of R_i can be chosen accordingly.
- *Resource-guided*: If the user has a resource constraint on the computation cost, memory or storage limit, the core tensors can also be adjusted based on that.

In the former case, the size of the core tensors may change depending on the streams' characteristics, while in the latter case, the reconstruction error may vary over time.

11.5 Performance Evaluation

Data Description SENSOR: The sensor data are collected from 52 MICA2 Mote sensors placed in a lab over a period of a month. Every 30 seconds each Mote sensor sends to the central collector via wireless radio four types of measurements: light intensity, humidity, temperature and battery voltage. In order to compare different types, we scale each type of measurement into zero mean and unit variance. This calibration process can actually be done online since mean and variance can be easily updated incrementally. Note that we put equal weight on all measurements across all nodes. However, other weighting schemes can be easily applied, based on the application.

Characteristics: The data are very bursty but still correlated across locations and time. For example, the measurements of same type behave similarly, since all the nodes are deployed in the same lab.

Tensor construction: By scanning through the data once, the tensor windows are incrementally constructed and processed/decomposed. More specifically, every tensor window is a 3-mode tensor with dimensions $(W, 52, 4)$ where W varies from 100 to 5000 in the experiments.

Parameters:

This experiment has three parameters: (1) *Window size* W: the number of time stamps included in each tensor window. (2) *Step ratio* S: the number of newly arrived tensors in the new tensor windows divided by window size W (a ratio between 0 and 1). (3) *Core tensor size*: (R_0, R_1, R_2) where R_0 is the size of time mode. IW and MW reach the same error level[3] across all the experiments, since we use the same termination criterion for the iterative algorithm in both cases.

Stable over time: Fig. 11.11(a) shows the CPU time as a function of elapsed time, where we set $W = 1000$, $S = .2$ (i.e. 20% new tensors). Overall, CPU time for both IW and MW exhibits a constant trend. MW achieves about 30% overall improvement compared to IW, on both datasets.

The performance gain of MW comes from its incremental initialization scheme. As shown in Fig. 11.11(b), the CPU time is strongly correlated with the number of iterations. As a result of MW, which reduces the number of iterations needed, MW is much faster than IW.

Consistent across different parameter settings: *Window size*: Fig. 11.12(a) shows CPU time (in log-scale) vs window size W. CPU time is increasing with window size. Note that the MW method achieves big computational saving across all sizes, compared to IW.

[3] Reconstruction error is $e(\mathcal{D}) = \frac{\|\mathcal{D}-\tilde{\mathcal{D}}\|_F^2}{\|\mathcal{D}\|_F^2}$, where the tensor reconstruction is $\tilde{\mathcal{D}} = \mathcal{D}\prod \times_i (\mathbf{U}_i\mathbf{U}_i^T)$.

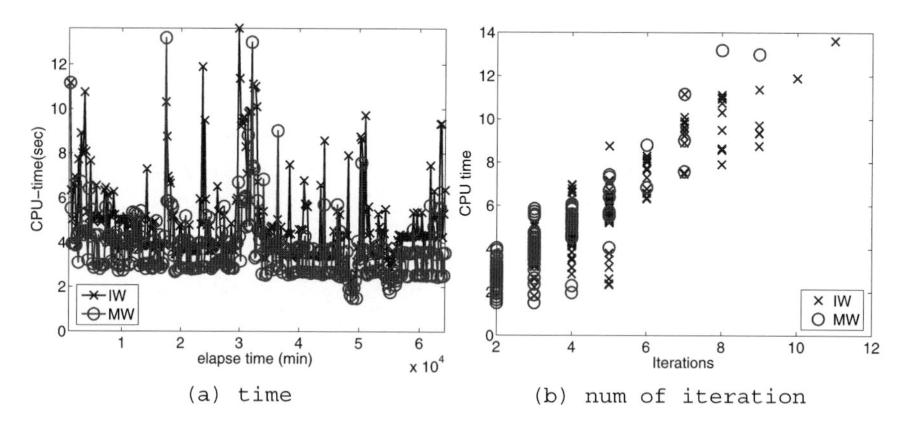

(a) time (b) num of iteration

Fig. 11.11 Time-evolving experiment

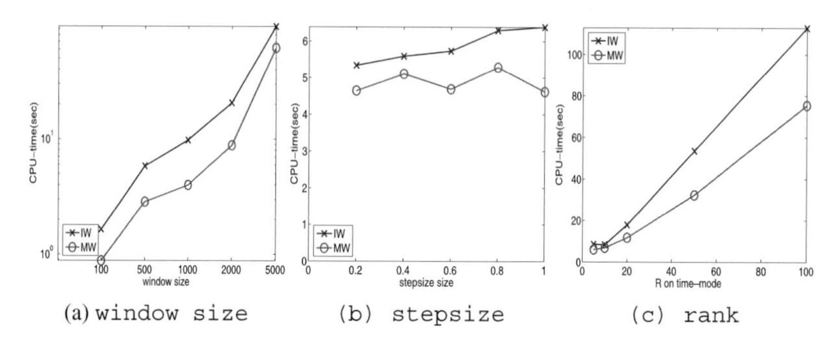

(a) window size (b) stepsize (c) rank

Fig. 11.12 Consistent over different parameters

Step size: Fig. 11.12(b) presents step size vs CPU time. MW is much faster than IW across all settings, even when there is no overlap between two consecutive tensor windows (i.e., step size equals 1). This clearly shows that the importance of a good initialization for the iterative algorithm.

Core tensor size: We vary the core tensor size along the time-mode and show CPU time as a function of this size (see Fig. 11.12(c)). Again, MW performs much faster than IW, over all sizes. Similar results are achieved when varying the sizes of the other modes, so we omit them for brevity.

11.6 Application and Case Study

In this section, we introduce a powerful mining application of window-based tensor analysis, *Multi-Aspect Correlation Analysis* (MACA). Then we present a case study of MACA on the SENSOR dataset. Figure 11.13 shows two factors across three aspects (modes): time, location and type. Intuitively, each factor corresponds to a trend

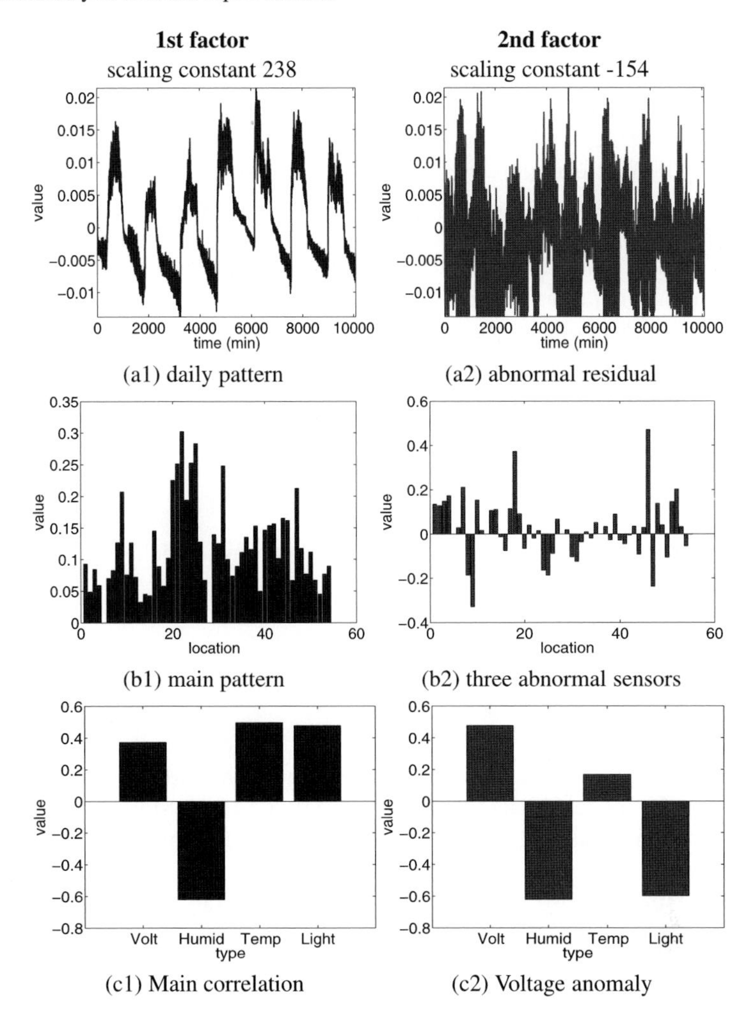

Fig. 11.13 Case study on environmental data

and consists of a global "importance" weight of this trend, a pattern across time summarizing the "typical behavior" for this trend and, finally, one set of weights for each aspect (representing their participation in this trend). More specifically, the trends are from projection matrices and the weights from core tensors.

Figures 11.13(a1, b1, c1) show the three components (one for each aspect) of the first factor, which is the main trend. If we read the components independently, Fig. 11.13 (a1) shows the daily periodic pattern over time (high activation during the day, low activation at night); (b1) shows the participation weights of all sensor locations, where the weights seem to be uniformly spread out; (c1) shows that light, temperature and voltage are positively correlated (all positive values) while they are anti-correlated with humidity (negative value). All of those patterns can be confirmed from the raw data. All three components act together scaled by the constant factor 238 to form an approximation of the data.

Part of the residual of that approximation is shown as the second factor in Figs. 11.13(a2, b2, c2), which corresponds to some major anomalies. More specifically, Fig. 11.13(a2) suggests that this pattern is uniform across all time, without a clear trend as in (a1); (b2) shows that two sensors have strong positive weights while one other sensor has a dominating low weight; (c2) tells us that this happened mainly at voltage which has the most prominent weight. The residual (i.e., the information remaining after the first factor is subtracted) is in turn approximated by combining all three components and scaling by constant -154. In layman terms, two sensors have abnormally low voltage compared to the rest; while one other sensor has unusually high voltage. Again all three cases are confirmed from the data.

11.7 Related Work

Next, we broadly review related work from the two major fields spanned by our work.

11.7.1 Data Streams

Data streams have been extensively studied in recent years. Recent surveys [19] have discussed many data stream algorithms. Among them, the sliding (or moving) window is a popular model in data stream literature [9,5,2,16,1]. Most of them monitor some statistics in the sliding window over a single stream using probabilistic counting techniques.

Multiple streams also have been studied in several places. For example, StatStream [28] uses the DFT to summarize streams within a finite window and then computes the highest pairwise correlations among all pairs of streams, at each time stamp. SPIRIT [20] performs PCA in a streaming fashion, discovering the hidden variables among the given n input streams and automatically determining when more or fewer hidden variables are needed. Atomic Wedgie [25] uses lower-bounding techniques to do subsequence matching of streaming time series to a set of predefined patterns.

All these methods work with a single stream or multiple streams of the same aspect, while we work with a more general notion of multi-aspect streams.

11.7.2 Tensor and Multi-Linear Analysis

Tensor algebra and multi-linear analysis have been applied successfully in many domains [6,14,24]. Powerful tools have been proposed, including the Tucker decomposition [23], parallel factor analysis [11] or canonical decomposition [4]. More recently, Ye [27] presented the generalized low rank approximation which extends

PCA from vectors (first-order tensors) into matrices (2nd order tensors). Ding and Ye [8] proposed an approximation of [27]. A similar approach is also proposed in [12]. Drineas and Mahoney [18] showed tensor CUR, applying it on a biomedical application. Kolda et al. [15] applied PARAFAC on web graphs to generalize the hub and authority scores for web ranking through term information.

These methods usually assume static data, while we are interested in streams of tensors. For the dynamic case, Sun et al. [22] proposed dynamic and streaming tensor analysis for higher-order data streams. However, [22] uses an exponential forgetting model instead of moving window, and more importantly, the time aspect is not analyzed explicitly as in this chapter. Using OLAP terminology, Hsieh et al. [13] presented algorithms for doing DCT and DWT on snapshots of data cubes over time, whereas our chapter focuses on tensor windows over time using the multilinear approach.

11.8 Conclusion

In collections of multiple, time-evolving data streams from a large number of sources, different data values are often associated with multiple aspects. Usually, the patterns change over time. In this chapter, we propose the *tensor stream* model to capture the structured dynamics in such collections of streams. Furthermore, we introduce *tensor windows*, which naturally generalize the sliding window model. Under this data model, two techniques, independent and moving window tensor analysis (IW and MW), are presented to incrementally summarize the tensor stream. The summary consists of local patterns over time, which formally correspond to core tensors with the associated projection matrices. Finally, extensive performance evaluation and case study demonstrate the efficiency and effectiveness of the proposed methods.

Acknowledgements We are pleased to acknowledge Brett Bader and Tamara Kolda from Sandia National lab for providing the tensor toolbox [3] which makes our implementation and experiments an easy job.

References

[1] A. Arasu, G.S. Manku, Approximate counts and quantiles over sliding windows. In: PODS, pp. 286–296, 2004.
[2] B. Babcock, M. Datar, R. Motwani, Sampling from a moving window over streaming data. In: SODA, pp. 633–634, 2002
[3] B.W. Bader, T.G. Kolda, Matlab tensor toolbox version 2.0, http://csmr.ca.sandia.gov/tgkolda/tensortoolbox/, 2006.
[4] J.D. Carroll, J. Chang, Analysis of individual differences in multidimensional scaling via an n-way generalization of 'eckart-young' decomposition. Psychometrika, 35(3), 1970.
[5] M. Datar, A. Gionis, P. Indyk, R. Motwani, Maintaining stream statistics over sliding windows: (extended abstract). In: SODA, pp. 635–644, 2002.

[6] L. de Lathauwer, Signal processing based on multilinear algebra. PhD thesis, Katholieke, University of Leuven, Belgium, 1997.
[7] A. Dempster, N. Laird, D. Rubin, Maximum-likelihood from incomplete data via the EM algorithm. Journal of the Royal Statistical Society B, 39(1):1–38, 1977.
[8] C. Ding, J. Ye, Two-dimensional singular value decomposition (2dsvd) for 2d maps and images. In: SDM, 2005.
[9] P.B. Gibbons, S. Tirthapura, Distributed streams algorithms for sliding windows. In: SPAA, pp. 63–72, 2002.
[10] G.H. Golub, C.F. Van-Loan, Matrix Computations, 2nd edn. The Johns Hopkins University Press, Baltimore, 1989.
[11] R. Harshman, Foundations of the parafac procedure: model and conditions for an "explanatory" multi-mode factor analysis. UCLA working papers in phonetics, 16, 1970.
[12] X. He, D. Cai, P. Niyogi, Tensor subspace analysis. In: NIPS, 2005.
[13] M.-J. Hsieh, M.-S. Chen, P.S. Yu, Integrating dct and dwt for approximating cube streams. In: CIKM, 2005.
[14] A. Kapteyn, H. Neudecker, T. Wansbeek, An approach to n-mode component analysis. Psychometrika, 51(2), 1986.
[15] T.G. Kolda, B.W. Bader, J.P. Kenny, Higher-order web link analysis using multilinear algebra. In: ICDM, 2005.
[16] L.K. Lee, H.F. Ting, Maintaining significant stream statistics over sliding windows. In: SODA, pp. 724–732, 2006.
[17] R.J. Little, D.B. Rubin, Statistical Analysis with Missing Data, 2nd edn. Wiley, New York, 2002.
[18] M.W. Mahoney, M. Maggioni, P. Drineas, M. w. mahoney, m. maggioni, and p. drineas. In: KDD, 2006.
[19] S. Muthukrishnan, Data Streams: Algorithms and Applications, vol. 1. Foundations and Trends in Theoretical Computer Science, 2005.
[20] S. Papadimitriou, J. Sun, C. Faloutsos, Streaming pattern discovery in multiple time-series. In: VLDB, 2005.
[21] J. Sun, S. Papadimitriou, C. Faloutsos, Distributed pattern discovery in multiple streams. In: Proceedings of the Pacific-Asia Conference on Knowledge Discovery and Data Mining (PAKDD), 2006.
[22] J. Sun, D. Tao, C. Faloutsos, Beyond streams and graphs: dynamic tensor analysis. In: KDD, 2006.
[23] L.R. Tucker, Some mathematical notes on three-mode factor analysis. Psychometrika, 31(3), 1966.
[24] N. Viereck, M. Dyrby, S.B. Engelsen, Monitoring Thermal Processes by NMR Technology. Elsevier, Amsterdam, 2006.
[25] L. Wei, E.J. Keogh, H.V. Herle, A. Mafra-Neto, Atomic wedgie: efficient query filtering for streaming times series. In: ICDM, 2005.
[26] F. Yates, The analysis of replicated experiments when the field results are incomplete. J. Exp. Agriculture, 1, 1933.
[27] J. Ye, Generalized low rank approximations of matrices. Machine Learning, 61, 2004.
[28] Y. Zhu, D. Shasha, Statstream: Statistical monitoring of thousands of data streams in real time. In: VLDB, 2002.

Part IV
Applications

Chapter 12
Knowledge Discovery from Sensor Data for Security Applications

Auroop R. Ganguly, Olufemi A. Omitaomu, and Randy M. Walker

Abstract Evolving threat situations in a post-9/11 world demand faster and more reliable decisions to thwart the adversary. One critical path to enhanced threat recognition is through online knowledge discovery based on dynamic, heterogeneous data available from strategically placed wide-area sensor networks. The knowledge discovery process needs to coordinate adaptive predictive analysis with real-time analysis and decision support systems. The ability to detect precursors and signatures of rare events and change from massive and disparate data in real time may require a paradigm shift in the science of knowledge discovery. This chapter describes a case study in the area of transportation security to describe both the key challenges, as well as the possible solutions, in this high-priority area. A suite of knowledge discovery tools developed for the purpose is described along with a discussion on future requirements.

Key words: Wide-area sensors, Heterogeneous data, Rare events, Knowledge discovery, Transportation security, Weigh stations

12.1 Introduction

Remote and in-situ sensors, including wireless sensor networks and large-scale or wide-area sensor infrastructures, as well as RFIDs, satellites and GPS systems [1,2,16], are becoming ubiquitous across multiple application domains ranging from disaster management [9], intelligence analysis and homeland or national security, to transportation, supply chain, ecology [10], climate change and the management of

A.R. Ganguly · O.A. Omitaomu · R.M. Walker
Computational Sciences and Engineering Division, Oak Ridge National Laboratory, 1 Bethel Valley Road, Oak Ridge, TN 37831 USA

A.R. Ganguly
e-mail: gangulyar@ornl.gov

critical infrastructures. On the one hand, massive volumes of disparate data, typically dimensioned by space and time, are being generated in real time or near real time. On the other hand, the need for faster and more reliable decisions is growing rapidly in the face of emerging challenges like natural, man-made, or technological failure-related disasters, intelligence analysis for security, competitive demand-driven logistics chains, ecological management in the face of environmental change, and the rapid globalization of economies and peoples. The critical challenge for researchers and practitioners is to generate actionable insights from massive, disparate and dynamic data, in real time or near real time.

12.2 Security Challenges

The need to anticipate the adversary and thwart terror attacks is a high-priority global problem. Terrorist attacks can range from improvised explosive devices (IED) and 9/11-type attacks on assets and critical infrastructures, as well as biological attacks potentially causing pandemic outbreaks, to the sabotage of power plants leading to chemical, biological and radiation pollutants in the atmosphere. A major area of concern is the possibility of terror outfits smuggling in materials for "dirty" nuclear bombs with the help of sleeper cells. This is a risk that demands urgent solutions as the potential impacts on human lives and on the economy defy comprehension. However, this battle needs to be continuously won on multiple fronts. These include border crossings, transit switching points like maritime ports, rail intermodal yards, and airports, as well as movement of cargo within the road or water transportation networks. The requirements range from the ability to distinguish between naturally occurring radioactive material (NORM), technologically enhanced NORM (TENORM), man-made radioactive materials in commerce, and special nuclear material (SNM) based on portal monitoring at border crossings to the need to assure multi-modal transportation security by preventing the movement of illicit radioactive materials along the links (e.g., roads, railway tracks, air and maritime routes) and nodes (e.g., inspection stations, train stations, airports and maritime ports) that comprise the complex global transportation network. The adversarial setting and the rarity of security-related violations makes the problem significantly more challenging. Thus, terrorists may shield unauthorized radioactive materials during transport, thus effectively rendering the radiation portal monitors useless. The problem of identifying anomalous cargo at the border crossings has also been investigated [3,7].

Wireless sensor networks and wide-area sensor infrastructures can generate large volumes of heterogeneous data from multiple types of sensors, ranging from video or image cameras to weather-observing instruments, as well as chemical, biological or radiation sensors. In addition, sensors can be used for tracking traffic movement, monitoring cargo in ships, airplanes or trucks, population displacements, and even significant change in certain types of social behavior. The massive volumes of disparate data generated from wide-area sensors in real-time lead to new challenges. However, sensor-based observations may point the way to solutions as well. Ely et al. [3] used two simple methods, ratio and slope measures, to compare the distributions of two NORM sources (cat litter and road salt) and two illegitimate sources (plutonium and highly enriched uranium) with the background. The ratio method normalizes the counts in the highest window and compares it to the corresponding background ratio. The slope method compares count rates in adjacent energy windows, forming a slope between them, to the corresponding background slope. In all their analyses, the effect of shielding was not investigated, which makes the focus of their problem very different from our definition of the problem. Let us consider the transport of shielded illicit radioactive materials, as referred to earlier. While sufficient shielding may cause one type of sensor, specifically radiation sensors, to have limited value, other types of sensor data, including but not limited to wide-area sensors that monitor US DOT numbers, license plates, cargo and axle weight, time of day/year, RFID tags, information about the carrier and the driver, or cargo loading profiles, may provide clues regarding suspicious activity or malicious intent. The other piece of the puzzle may be computer-based simulations generated from, for example, computational social scientific or dynamical process-oriented models. The computational models may generate emergent behavior and predictive insights at multiple space-time scales. An intelligent combination of model-simulations and sensor measurements can pave the way for enhanced cognizance and mitigation of threat scenarios.

12.3 Disparate Data Exploitation

While massive, dynamic and disparate data may provide a pathway to solution strategies for security challenges, the data also represent a challenge in themselves. We are said to live in an information age and in the middle of a data explosion. However, this data explosion is expected to grow worse as sensors and sensor networks start becoming ubiquitous and as they improve upon their ability to gather disparate and high-resolution data, which may be dimensioned by space and time, in real time. Our ability to exploit the vast quantities of disparate data for high-priority application solutions will depend critically on how the data is processed and used. The following provides a high-level overview of the three major steps for the data exploitation process:

1. From Data to Information via Data Fusion: Raw data need to be converted to information, through systematic collection, integration, organization, presentation and summarization. The process has been referred as data fusion. Sensor-based data fusion is an area of active research, and there are existing definitions of data fusion levels which go all the way from pure data integration to generating useful insights.
2. From Information to Knowledge via Knowledge Discovery: The process of converting information to actionable insights through adaptive predictive analysis has been defined as knowledge discovery. In an effort to differentiate non-trivial discovery from basic statistical trend analysis, the research community has often defined knowledge discovery as the ability to generate non-trivial, non-intuitive and meaningful insights from information.

3. From Knowledge to Decisions via Decision Support: Information obtained from data fusion and the knowledge gained from offline discovery need to be used in conjunction with real-time analysis for decision-making. The requirements may range from automated time-phased or event-based recommendations in real time or near real time to facilitating end-user decisions for tactical, operational and strategic requirements.

The three-step process above typically needs to be deeply tied to the context of the domain or the application where the information, knowledge or decisions are sought. However, commonalities do emerge in the requirements as well as the tools that need to be developed for data fusion, knowledge discovery and decision support. The area of data fusion has been studied in depth. See Llinas and Waltz [8], Hall and Llinas [5], and Klein [6] for a review of data-fusion techniques. Here we discuss the broad requirements in three emerging and overlapping areas that are relevant for disparate data exploitation, over and above traditional data fusion: *data sciences*, *computational sciences* and *decision sciences*.

12.3.1 Data Sciences

A new focus may be needed within the data sciences to handle the tremendous challenges encountered in security applications. Although data mining algorithms often, but not always, assume independence of learning samples and envisage scenarios that require either supervised or unsupervised learning, the community has developed tools and methods for mining from dependent samples as in from temporal, spatial or spatio-temporal data, as well as certain semi-supervised learning paradigms where firm labels are absent but crude domain knowledge may be present. However, further developments are required in these relatively new areas, especially in areas like spatio-temporal data mining and semi-supervised learning when the domain knowledge present is known to be useful, but cannot be easily translated to simple rules or embedded within learning examples. Similarly, the area of anomaly detection from massive and disparate data, both in online and offline modes, has received considerable attention in the last couple of years by the data mining community. However, much remains to be done in terms of automated extraction and detection of anomalies as well as in developing leading indicators, which can either predict such anomalies or use the anomalies as predictors depending on the context. Such leading indicator analysis would have to occur in an automated mode, especially with massive and heterogeneous data dimensioned by space and time. There are a few other significant methodological areas which are critical for security applications but have so far received limited attention from the data mining community. We pick two examples for illustration, specifically, nonlinear statistics and extreme value theory. Nonlinear statistics has been traditionally explored by a specialized group within the physics community while extreme value theory is a focus area within statistics. However, pressing societal concerns ranging from natural disasters, climate change, terrorism and pandemics have, over the last few years, brought these relatively arcane disciplines to the forefront of science and technology. The new challenges require non-traditional and new data sciences spanning traditional disciplines like statistics, computer science, nonlinear dynamics, signal processing and econometrics.

12.3.2 Computational Sciences

The requirements for sensor information processing vary from archived analysis or *offline knowledge discovery* within a central computing, perhaps supercomputing, environments, to real-time or near-real-time analysis or *online knowledge discovery* and event-based processing, within an in-situ or remote unit, often in a distributed processing environment. A distinction among the two modes of operation may arise as a result of the processing and data-integration requirements or the speed and scalability of the data science algorithms. The applicability of non-traditional and novel algorithms to massive and disparate data dimensioned by space and time remain of interest for both situations. The computational science aspects that can be of significant interest to the data science community are what may be called *computational efficiency* and *algorithmic efficiency*. We use the former phrase in this context to include innovations in computational speeds, data structures, processing and scalability, primarily for offline processing. The latter phrase is used to describe the development of the mathematical algorithms themselves in a manner such that they are fast, storage- and memory-efficient, adaptive, and work in real time or near real time for robust online implementations. In addition, the relevant challenges in the computational sciences include data integration and fusion as well as coordination between offline and online discovery, besides scaling up to high-performance modeling and simulation environments.

12.3.3 Decision Sciences

The decision scientific requirements range from strategic policy-making, for example, in the context of advance disaster planning and readiness efforts, to real-time decision aids, for example, to detect, thwart and/or manage disasters. The decision phase, and hence the decision scientific requirements, is typically strongly coupled to the specific application domain. In addition, this is the phase that brings the "human in the loop" in the form of analysts, decision-makers and policy strategists. The primary requirements for which advanced tools need to be developed or refined can be categorized into the following: *hypothesis generation* and *decision support*. The hypothesis generation, or abduction, phase includes methods for advanced visualizations and analysis induced through human-computer interactions. In addition to human-driven hypothesis generation, emerging areas like automated generation of hypothesis or causal models, as well as new advances in artificial intelligence, may need to be explored. The decision support aspect comprises traditional tools like on-line analytical processing and geographic information services, with embedded data scientific methodologies, coordination with tools in areas like risk analysis and operations research, as well as the development of efficient metrics and tools that can aid tactical, operational or strategic decision-making.

12.4 Case Study: Weigh Station Sensors

12.4.1 Domain Description

The ability to ensure inter-modal transportation security, specifically to deter the transport of contraband materials or unauthorized movement of people, is a requirement that exists at the borders of nations as well as at defense or security checkpoints and inspection stations. The inspection stations can be placed at each link and node of the inter-modal transportation network, including maritime ports and airports as well as rail stations and truck weigh stations. Here we consider the movement of cargo along the road transportation networks in the hinterland of ports, where contraband radioactive cargo transported in ships to the sea ports of a nation evade port security checkpoints and attempt to make their way via the road networks to vulnerable and high-value destinations. As a specific example, we consider the port of Mobile in Alabama. The average time lines for transporting cargo on the highway from the port to interior points in the US has been shown on the transportation network in Fig. 12.1a. As shown in the figure, within just six to eight hours the coverage includes the important cities of Atlanta, Memphis, Houston, and New Orleans, while within a bit more than half a day's travel the coverage spans several other important cities including Washington, DC, and Chicago. In less than a day of highway travel time, the coverage spans the entire South and Middle Atlantic regions and most of the Midwest, Southwest, and New England regions of the United States.

A similar average time lines for transporting cargo from the same port on the railway is shown in Fig. 12.2. It is evident from this figure that in less than half a day of rail travel time, the coverage spans the entire South region and a part of the Midwest region of the United States.

Thus, a terrorist group that can successfully evade the port security system with illicit radioactive materials may be able to successfully transport the materials along the US road transportation networks to most high-value targets on the eastern half of the country within a day. This rather frightening scenario has caused national and homeland security experts in the US to discuss the possibility of providing secure transportation corridors where safe and secure transportation of cargo can be assured with a degree of confidence from the ports to sensitive and vulnerable destinations.

Our specific interests are to ensure that illicit radioactive materials, specifically materials for "dirty bombs", can be detected as they are moved along the secure

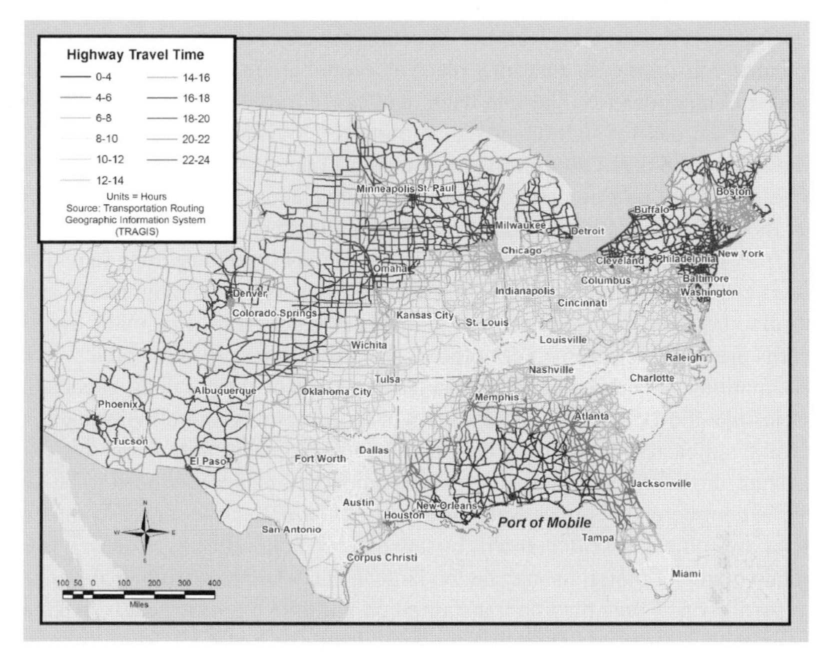

Fig. 12.1 Highway travel times for cargo movement from Mobile, AL

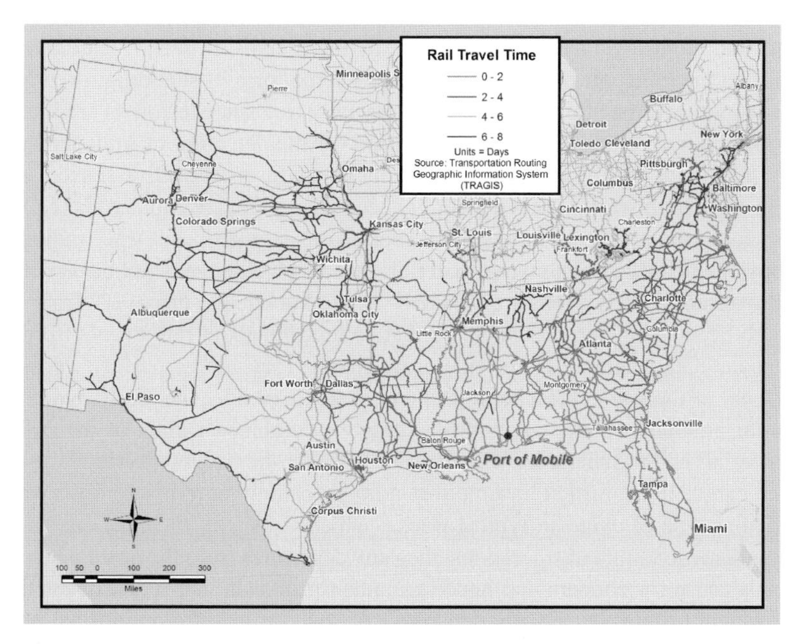

Fig. 12.2 Rail Travel Times for Cargo Movement from Mobile, AL

transportation corridor. However, the detection process cannot be excessively time or resource consuming, as that may place a severe burden on the law enforcement budgets and manpower on the one hand, and cause traffic delays and in the worst case may even end up significantly disrupting the logistics chain and commerce. On the other hand, even though the movement of contraband radioactive materials may be extremely rare compared to the total movement of goods, the inability to detect those rare events may lead to catastrophic disasters. In addition, the decision on whether to perform detailed manual inspection may need to be done in real time. Thus, a key requirement is to ensure that decision makers have the right tools at their disposal to be able to make faster and more reliable decisions. Pragmatic and legal concerns prevent law enforcement officers from searching private vehicles. In fact, an assumption can perhaps be made that terror groups or sleeper cells, especially in countries like the US, do not possess enough resources to acquire high-quality and expensive radioactive materials which may be used to cause a dirty nuclear explosion. Thus, the relatively larger volumes of radioactive cargo may need to be transported in commercial vehicles, specifically trucks, which move cargo. Commercial vehicles are subject to search at will by law enforcement officers. However, even if the total volume of materials may be too large to carry within private vehicles, there is a possibility that they may be transported on multiple trucks in relatively smaller amounts with appropriate shielding to camouflage the materials from radioactive sensors. The possibility of relatively small terror cells and sleeper groups getting access to multiple trucks is relatively low, however, thus the assumption that relatively large volumes of materials need to be placed within one truck appears reasonable. The assumptions stated earlier lead to an interesting consideration. Even though the primary interest may be to detect illicit radioactive materials, the use of radioactive sensors and monitors may not be adequate by themselves. In fact, it is quite likely that during the transport of the illicit materials some form of shielding, even if in a crude fashion, will be attempted. This is especially true since terrorists are likely to be aware that radiation monitors may be used for the detection of radioactive materials. The crude shielding may be sufficient to confuse radioactive sensors and a law enforcement inspector. Since naturally occurring radioactive materials (NORM) like bricks, ceramic and cat litter are transported on a regular basis, trace radioactivity from shielded illicit radioactive materials that are shielded may not be detectable. The above fairly likely scenario is exactly where knowledge discovery from data, or the ability to discover actionable predictive insights about rare events from heterogeneous data available from wide-area sensors, may be the perfect weapon in the arsenal of the law enforcement decision maker. Consider the likely hypothesis, given our previously stated assumptions, that the combined weight of the radioactive materials and the shielding will significantly alter the weight profile of the truck and the cargo. If we further assume that normal loading patterns are observed for a particular type of cargo in typical trucks, which is usually the case due to various pragmatic loading considerations, then any departures from the normal behavior may be a cause for concern and hence grounds for manual inspection of the truck. Thus, weight profiles, especially departures of such profiles from a normal behavior, may indirectly help in the detection of illicit radioactive materials which may be camouflaged from radiation sensors due to shielding. This holds especially because shielding can be achieved with materials like lead. Weight profiles can indeed be obtained for trucks from weighing scales. In fact, truck weigh stations across the US and other countries often monitor the weight of trucks anyway to ensure that they follow weights and measures laws and maintain safe vehicles and drivers; for example, in terms of maintaining a maximum allowable weight-to-length ratio depending on the type of vehicle. Other types of data from wide area sensors, for example, manifest information which carries information about the expected movement and cargo of the truck, license plate number, driver licenses, and other truck or driver identification information, can all help narrow down the search for trucks that represent plausible security hazards. The wealth of data which can be gathered for trucks is shown in Fig. 12.3.

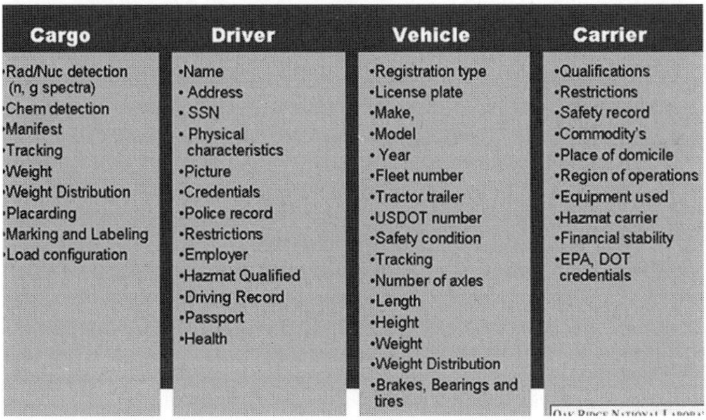

Cargo	Driver	Vehicle	Carrier
•Rad/Nuc detection (n, g spectra)	•Name	•Registration type	•Qualifications
•Chem detection	• Address	•License plate	•Restrictions
•Manifest	• SSN	•Make,	•Safety record
•Tracking	• Physical characteristics	•Model	•Commodity's
•Weight	•Picture	• Year	•Place of domicile
•Weight Distribution	•Credentials	•Fleet number	•Region of operations
•Placarding	•Police record	•Tractor trailer	•Equipment used
•Marking and Labeling	•Restrictions	•USDOT number	•Hazmat carrier
•Load configuration	•Employer	•Safety condition	•Financial stability
	•Hazmat Qualified	•Tracking	•EPA, DOT credentials
	•Driving Record	•Number of axles	
	•Passport	•Length	
	•Health	•Height	
		•Weight	
		•Weight Distribution	
		•Brakes, Bearings and tires	

Fig. 12.3 Heterogeneous data gathering requirements for motor carriers

12.4.2 Weigh Station Test-bed Description

The SensorNet®program managed by the Computational Sciences and Engineering Division of the Oak Ridge National Laboratory has developed test beds to study the ability of sensor-based data to detect trucks that represent plausible security or safety hazards. The test beds have been developed in existing truck weigh stations at interstates. Wide-area sensors have been placed at these test beds, where the term "sensor" is rather broadly construed to include the types of detectors indicated in Figs. 12.4 and 12.5.

Here we focus on one test bed, specifically the weigh station at the I-40 intersection near Watt Road in Tennessee, USA. The process of truck inspections at weigh stations with sensor test beds is depicted in Fig. 12.6.

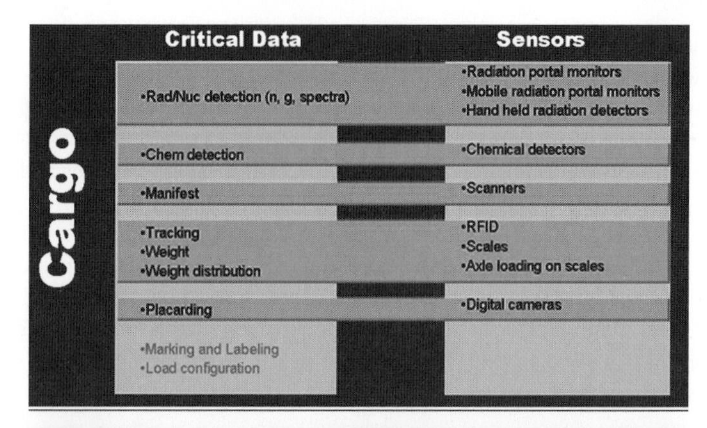

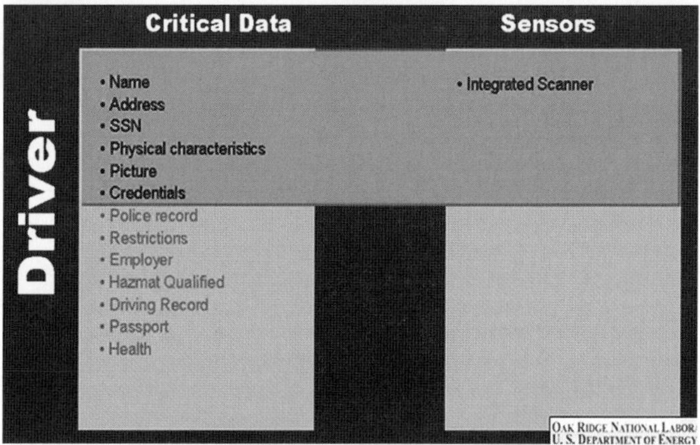

Fig. 12.4 Some critical heterogeneous data captured by sensors at weigh station test beds

12.4.3 Knowledge Discovery Process

Our objective is to demonstrate that heterogeneous data from historical truck records can be used to generate patterns and actionable insights regarding the normal behavior of trucks, which in turn can be used for mining anomalies in the data, and subsequently coordinated with event-based analysis of real-time truck data for online detection of anomalies, abnormalities or change. The ultimate aim is to facilitate faster and more reliable end-user decisions on whether a new truck at the weigh station represents a plausible security or safety hazard and hence should be held for manual inspection. The state of the art is depicted in Fig. 12.7, which shows an example of how the information for weigh station trucks can be currently visualized by end-users and decision makers.

We have developed a suite of knowledge discovery approaches to facilitate the decision-making process. The approaches include linear and nonlinear correlation estimation from short and noisy data [1]; offline pattern detection and anomaly

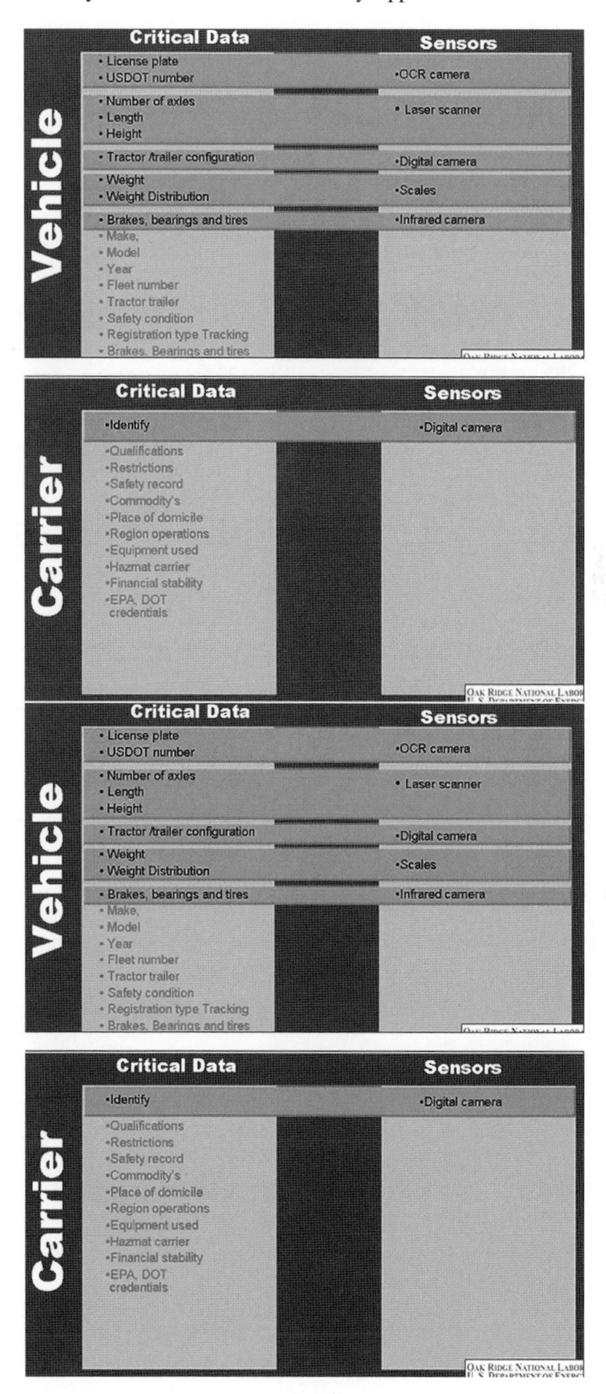

Fig. 12.5 Other critical heterogeneous data captured by sensors at weigh station test beds

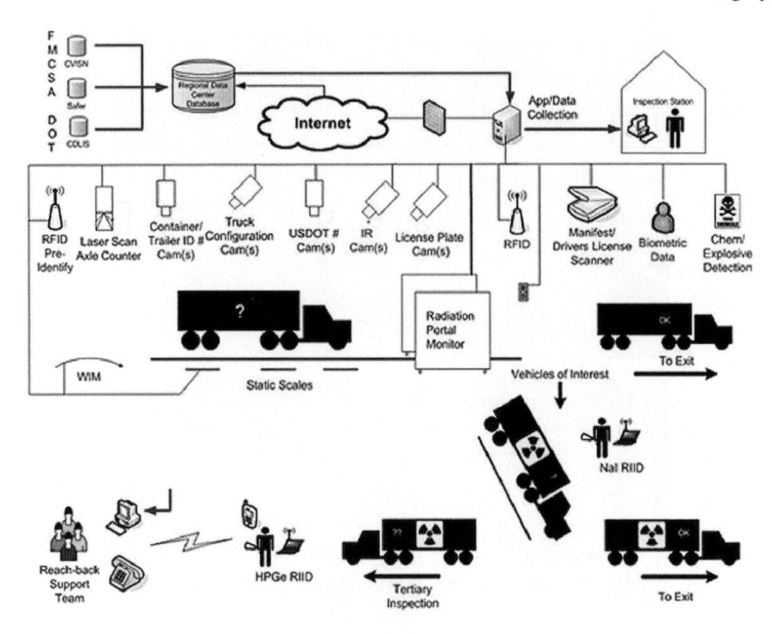

Fig. 12.6 Truck inspection process flow diagram at weigh station test beds

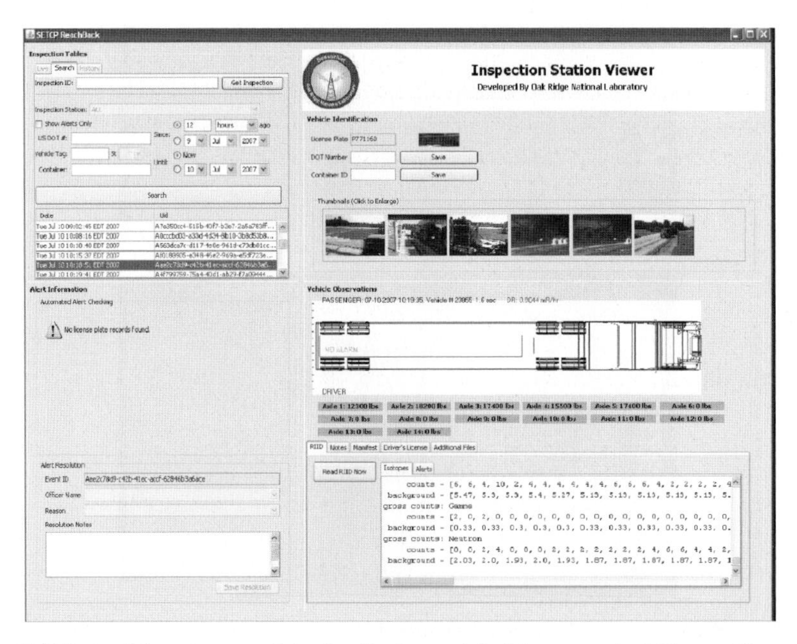

Fig. 12.7 State-of-the-art sensor data visualization and decision support platform at the weigh station test beds

analysis based on linear and nonlinear dimensionality reduction [1]; probabilistic dimensionality reduction for online anomaly detection [4]; extreme-value theory and profile decomposition approaches combined with effective distance measures for profile matching and clustering [12,14]; wavelet methods for reducing random fluctuations or inherent noise in data sets [11,13]; goodness-of-fit approaches for classification of profiles and deviation detection, as well as online change detection from dynamic and evolving data. The approaches have been, and are being, validated with surrogate data as well as with real observations from weigh station sensors and other related applications. Based on this suite of approaches, we have also developed a proof-of-concept decision user interface (UI). The interface, designed for decision makers, showcases capabilities for real-time analysis in the context of decision making. Specifically, analyses based on static scale data, gross radiation counts as well as radiation spectroscopy data are demonstrated. Each set of analysis has a corresponding offline knowledge discovery component, which in turn can influence the online analysis and real-time decision aspects. While the specific tools and processes have been described elsewhere or will be described in later publications, we discuss three different scenarios that have been used for demonstration purposes:

Scenario 1: Consider a truck of known weight and length that approaches the Watt Road weigh station with unknown cargo from Crossville, TN. The law enforcement needs to determine, within a few seconds, if this truck should be pulled out for closer inspection. The state-of-the-art method (shown in Fig. 12.7) generates an alarm based on the sensor gross data; therefore, it is required that the truck be manually inspected as explained in Sect. 4.2. A closer look at the truck shows that the cargo is an example of NORM, which ordinarily will generate alarm because of the presence of naturally occurring radioactive materials in the cargo. Such nuisance alarms disrupt supply chain networks and create additional overhead costs for weigh station administration. The use of our knowledge discovery (KD) based methodology shows that this truck should be dispatched without manual inspection. A proof of concept decision user interface for this scenario is shown in Fig. 12.8. The figure shows that the incoming truck data (shown in red for all plots) is below a threshold set for the static scale data using the truck weight and speed the two left plots. The use of truck weight and speed for this analysis is a result of an offline analysis of seven static scale data parameters. The results of the offline analysis show that these two parameters contain most information about the trucks. The offline analysis is based on linear and nonlinear dimensionality reduction approaches such as multi-dimensional scaling (MDS), locally linear embedding (LLE), and isometric feature mapping (ISOMAP). Agovic et al. [1] presented an application of these approaches for transportation security. Figure 12.8 shows further that the mean of the gross count is within the lower and upper confidence bounds for the so-called norm database (the two middle plots). The mean profile, upper and lower confidence bounds are also based on the results of the offline analysis. The spectroscopy profile analysis (last two plots) indicates that the cargo spectroscopy profile is within the norm for some commonly known cargoes.

The offline analysis usually takes time and may not be useful for real-time decision. The issue of scalability has become a serious issue in the data mining commu-

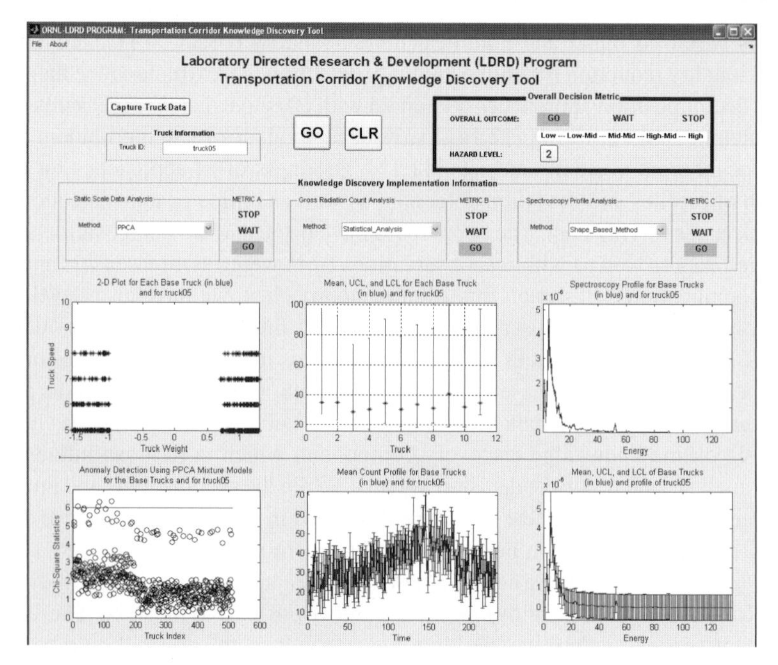

Fig. 12.8 Proof-of-concept user interface for online knowledge discovery and real-time decisions for Scenario 1

nity especially for online knowledge discovery applications in which decisions are expected in real time or near real time. The three categories of techniques for improving scalability in data mining are designing a faster algorithm, using a more efficient data representation, and partitioning the data [15], which has been called incremental or online knowledge discovery in data mining literature. In this research, the online knowledge discovery approach is implemented. Therefore, the offline analysis can be implemented in the background and the results fed into the online component, which is then used for real-time decision making. The online parameters and variables are updated regularly based on the results of the offline analysis.

Scenario 2: In this case, we consider another truck loaded with an unknown cargo from Mobile, AL. The UI snapshot for this scenario is shown in Fig. 12.9. These analyses give some interesting insights. Both the static scale and the gross data analyses give a STOP metric, but the spectroscopy analysis gives a GO metric. These results mean either of two things: there may be harmful material in the cargo but the radiation sensor is not detecting it because the material is well shielded; on the other hand, the shielding material shows up as additional weight in the static scale data or this truck may be violating size and weight regulations. In either case, the overall metric is "WAIT", which means that additional information about the truck should help in the final decision. Such additional information may lead to a secondary inspection.

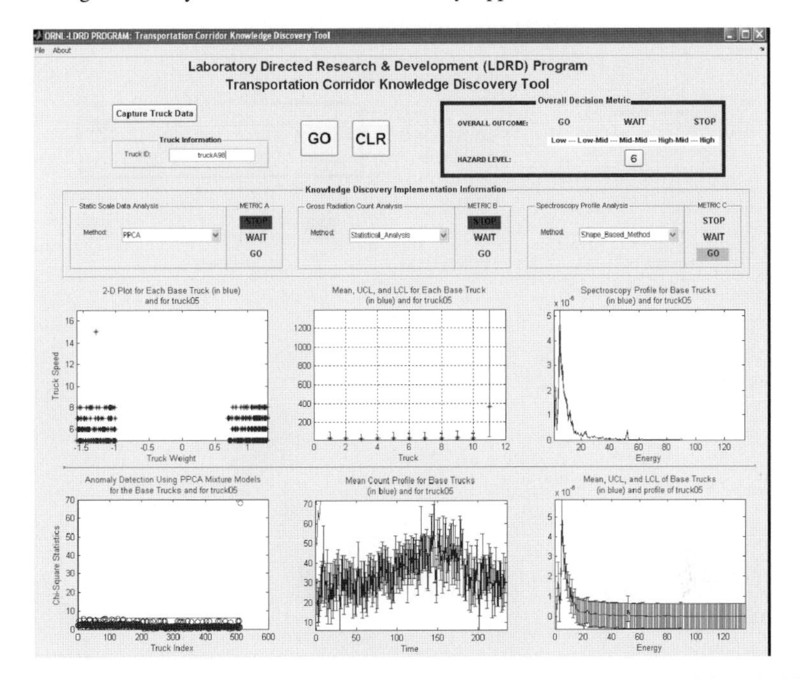

Fig. 12.9 Proof-of-concept user interface for online knowledge discovery and real-time decisions for Scenario 2

Scenario 3: This scenario considers a truck loaded with an unknown cargo from Little Rock, AR. For this truck, our KD-based tool generates STOP overall metric. The UI for this scenario is shown in Fig. 12.10. The overall metric is a STOP because both the gross count and the spectroscopy profile analyses have higher weight for computing the overall decision metric than the static scale result. This is a case in which the material may not be sufficiently shielded so that the radiation sensor can still pick up the radioactivity in the cargo. Since the material is not shielded or not shielded so much, the extra weight does not show up in the static scale data.

12.5 Closing Remarks

The ability to ensure inter-modal transportation security, especially as it relates to movement of cargo along the road transportation networks in the country, has become a national priority. The specific objective of this effort is to ensure that illicit radioactive materials, even if shielded or camouflaged, can be detected if and when they are moved along the secure transportation corridors. The detection process, which is based on knowledge discovery from heterogeneous sensor data at weigh stations, needs to be reliable and fast. The societal cost of missing the rare events (where an event is a new truck entering a weigh station and a rare event is a truck

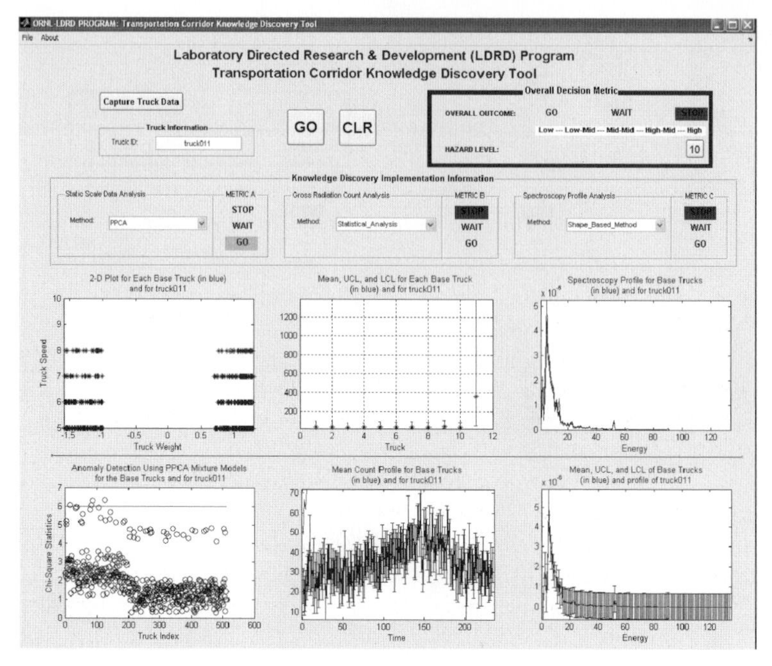

Fig. 12.10 Proof-of-concept user interface for online knowledge discovery and real-time decisions for Scenario 3

carrying illicit radioactive material) can be too high to compute, while excessive nuisance alarms or delays in the decision process may result not only in disrupting commerce but also perhaps permanently eroding the confidence of the decision maker in the capabilities of the automated tools to provide meaningful guidance. The latter can be dangerous as the purpose of the automated tools is to aid the human law enforcement officer as (s)he makes a decision on the need for manual inspections. In this chapter, we have presented knowledge discovery approaches for detecting anomalous trucks carrying illicit radioactive materials using truck weigh station sensors data: static scale data, cargo gross count, and cargo spectroscopy data. The suite of offline and online knowledge discovery approaches (discussed in other publications) are based on linear and nonlinear correlation estimation, probabilistic dimensionality reduction, extreme value theory and profile decomposition, wavelet denoising methods, and goodness-of-fit approaches for profile classification and deviation detection. The overall approach and decision metrics rely on anomaly detection and decision-making based on each type of data source as well as on a combination of the heterogeneous data sources. A proof of concept user interface developed based on this suite of approaches was presented in this chapter. Three different scenarios were discussed to demonstrate the application of the decision user interface and showcase its capabilities for real-time analysis. Future plans in the application area include extensive field testing of the methodologies and the decision user interfaces for the weigh station application as well as extension of the capabilities of the interface to other potential and related applications, for example, border crossings, railway stations and ports. The anomaly detection methodologies can be generalized for rare event or change detection in other application domains. Our broad anomaly detection approach has been to establish patterns of normal behavior; refine and build on the patterns as more and more use cases become available; flag the deviations from the normal as potential anomalies with a metric that quantifies the degree of departure; and hence provide guidance on when human intervention is required and can be useful. The advantages of this approach is that the patterns of normal behavior, and hence the anomaly detection approaches, are expected to grow stronger over time. Also, application-specific insights can be combined with data-dictated insights; advanced semi-supervised learning schemes can be leveraged or developed to incorporate domain-specific rules or insights; and the use of cases accumulated through experience or training from data can be utilized to continually improve decisions. Indeed, we believe this new kind of knowledge discovery approaches will be adopted more and more in applications like national or homeland security where rare but high-impacts events must be detected, specifically under situations where user- or application-specific knowledge may be limited but need to be used and can add value when they are indeed available, the raw data is heterogeneous and potentially massive, the number of cases or events are very large but the cases representing the rare events that need to be detected are either too few or perhaps even non-existent.

Acknowledgements This research was funded by the Laboratory Directed Research and Development (LDRD) Program of the Oak Ridge National Laboratory (ORNL), managed by UT-Battelle, LLC, for the U.S. Department of Energy under Contract DE-AC05-00OR22725. We would like to gratefully acknowledge synergistic research activities by the SensorNet®Program managed by the Computational Sciences and Engineering Division of the Oak Ridge National Laboratory. We specifically thank the following people in the SensorNet®group: Frank DeNap, Cy Smith, David Hill, and David Feaker. We also thank Brian Worley, Vladimir Protopopescu, Bruce Patton, Ian Gross, and Steven Saavedra for their individual and collective contributions to the Transportation Corridor research. We are thankful to Budhendra Bhaduri, Mark Tuttle, Cheng Liu, and Paul Johnson for providing the highway and rail travel time plots used in Figs. 12.1 and 12.2 respectively. We are also thankful to our internal technical reviewers, Cheng Liu and Raju Vatsavai, for reviewing this manuscript and their respective suggestions in improving the quality of the manuscript. We also acknowledge the use of US DOE's TRAGIS system developed at ORNL. This manuscript has been authored by UT-Battelle, LLC, under contract DE-AC05-00OR22725 with the U.S. Department of Energy. The United States Government retains and the publisher, by accepting the article for publication, acknowledges that the United States Government retains a non-exclusive, paid-up, irrevocable, world-wide license to publish or reproduce the published form of this manuscript, or allow others to do so, for United States Government purposes.

References

[1] A. Agovic, A. Banerjee, A.R. Ganguly, V.A. Protopopescu, Anomaly detection in transportation corridors using manifold embedding. In: Proceedings of the First International Workshop on Knowledge Discovery from Sensor Data, ACM KDD Conference, San Jose, CA.

[2] S.P. Chong, C.-Y. Kumar, Sensor networks: evolution, opportunities, and challenges. Proceedings of the IEEE, 91(8):1247–1256, 2003.
[3] J.H. Ely, R.T. Kouzes, B.D. Geelhood, J.E. Schweppe, R.A. Warner, Discrimination of naturally occurring radioactive material in plastic scintillator materials. IEEE Transactions on Nuclear Science, 51(4):1672–1676, 2004.
[4] Y. Fang, A.R. Ganguly, Mixtures of probabilistic principal component analyzers for anomaly detection. In: Proceedings of the First International Workshop on Knowledge Discovery from Sensor Data, ACM KDD Conference, San Jose, CA.
[5] D.L. Hall, J. Llinas, An introduction to multisensor data fusion. Proceedings of the IEEE, 85(1):6–23, 1997.
[6] L.A. Klein, Sensor and data fusion: a tool for information assessment and decision making. Society of Photo-Optical Instrumentation Engineering (SPIE) Press Monograph, PM138, 2004.
[7] R.T. Kouzes, J.H. Ely, B.D. Geelhood, R.R. Hansen, E.A. Lepel, J.E. Schweppe, E.R. Siciliano, D.J. Strom, R.A. Warner, Naturally occurring radioactive materials and medical isotopes at border crossings. IEEE Nuclear Science Symposium Conference Record, 2:1448–1452, 2003.
[8] J. Llinas, E. Waltz, Multisensor Data Fusion. Artech House, Boston, 1990.
[9] K. Lorincz, D.J. Malan, T.R.F. Fulford-Jones, A. Nawoj, A. Clavel, V. Shnayder, G. Mainland, M. Welsh, Sensor networks for emergency response: challenges and opportunities. Pervasive Computing, 16–23 October–December 2004.
[10] A. Mainwaring, J. Polastre, R. Szewczyk, D. Culler, J. Anderson, Wireless sensor networks for habitat monitoring. In: Proceedings of the 1st ACM International Workshop on Wireless Sensor Networks and Applications, pp. 88–97, 2002.
[11] O.A. Omitaomu, A.R. Ganguly, V.A. Protopopescu, Statistical analysis and wavelet-based denoising methods for analyzing truck radiation data. Oak Ridge National Laboratory, Technical Report: ORNL/TM-2006/596, 2006.
[12] O.A. Omitaomu, A.R. Ganguly, V.A. Protopopescu, A methodology for real-time decisions in sampling of trucks based on online anomaly analysis and radiation sensor data. Oak Ridge National Laboratory, Technical Report: ORNL/TM-2006/602, 2006.
[13] O.A. Omitaomu, A.R. Ganguly, V.A. Protopopescu, Denoising Radiation Sensor Data for Transportation Security Applications, working paper, 2007.
[14] O.A. Omitaomu, A.R. Ganguly, B.W. Patton, V.A. Protopopescu, Hierarchical clustering approach for anomaly detection in radiation sensor data. Working paper, 2007.
[15] F. Provost, V. Kolluri, A survey of methods for scaling up inductive algorithms. Data Mining and Knowledge Discovery, 3:131–169, 1999.
[16] M. Tubaishat, S. Madria, Sensor networks: an overview. IEEE Potentials, 20–23 April/May 2003.

Chapter 13
Knowledge Discovery from Sensor Data For Scientific Applications

Auroop R. Ganguly, Olufemi A. Omitaomu, Yi Fang, Shiraj Khan, and Budhendra L. Bhaduri

Abstract The current advances in sensors and sensor infrastructures offer new opportunities for monitoring the operations and conditions of man-made and natural environments. The ability to generate insights or new knowledge from sensor data is critical for many high-priority scientific applications especially weather, climate, and associated natural hazards. One example is sensor-based early warning systems for geophysical extremes such as tsunamis or extreme rainfall, which can help preempt disaster damage. Indeed, the loss of life during the 2004 Indian Ocean tsunami may have been significantly reduced, if not totally prevented, had sensor-based early warning systems been in place. One other example is high-resolution risk-mapping of insights obtained through a combination of historical and real-time sensor data, with physics-based computer simulations. Weather, climate and associated natural hazards have established history of using sensor data, such as data from DOPPLER radars. Recent advances in sensor technology and computational strengths have created a need for new approaches to analyzing data associated with weather, climate, and associated natural hazards. Knowledge discovery offers tools for extracting new, useful and hidden insights from data repositories. However, knowledge discovery techniques need to be geared towards scalable and efficient implementations of predictive insights, online or fast real-time analysis of incremental information, and solution processes for strategic and tactical decisions. Predictive insights regarding weather, climate and associated natural hazards may require models of rare, anomalous and extreme events, nonlinear phenomena, and change analysis, in particular from massive volumes of dynamic data streams. On the other hand, historical data may also be noisy and incomplete, thus robust tools

A.R. Ganguly · O.A. Omitaomu · Y. Fang · S. Khan · B.L. Bhaduri
Computational Sciences and Engineering Division, Oak Ridge National Laboratory, 1 Bethel Valley Road, Oak Ridge, TN 37831 USA

A.R. Ganguly
e-mail: gangulyar@ornl.gov

need to be developed for these situations. This chapter describes some of the research challenges of knowledge discovery from sensor data for weather, climate and associated natural hazard applications and summarizes our approach towards addressing these challenges.

Key words: Sensors, Knowledge discovery, Scientific applications, Weather extremes, Natural hazards

13.1 Introduction

Predictive insights generated from sensor data, in conjunction with data obtained from other sources like computer-based simulations, can facilitate short-term decisions and longer-term policies. Remote sensors [35], such as Earth-observing satellites, weather radars, large-scale sensor infrastructures [38] and environmental wireless sensor networks [5], yield massive volumes of dynamic and geographically distributed sensor data at multiple space-time resolutions. We define sensors broadly to include wireless sensor networks, in-situ sensor infrastructures and remote sensors. The raw data need to be converted to summary information and subsequently used to generate new knowledge or insights, ultimately leading to faster and more accurate tactical and strategic decisions. Therefore, we define knowledge discovery as the overall process where raw data from sensors or simulations are ultimately converted to actionable predictive insights for decision and policy makers. In addition to observations, scientific applications demand that information about the known physics, or data-dictated process dynamics, be taken into account. The scientific domains are diverse and requirements for sensor-based data processing and analysis can be fairly broad on one hand and domain specific on the other. This chapter focuses on applications of knowledge discovery from sensor data for weather, climate and geophysical hazards; these applications may be useful for hazards mitigation [17]. However, we present a broader view of knowledge discovery as compared to the traditional definitions by the data mining community, but we include the data mining and other data sciences as key aspects of the overall process.

Hazards can be natural [50], such as weather extremes including rainfall, hurricanes and heat waves; they can be technological, such as leakage and spread of toxic plumes from industrial facilities [23]; or they can be adversarial, as in security [1] and war. This chapter focuses primarily on hazards due to weather or climate extremes [33]. The idea of using intelligent data sciences and sensor data for hazards mitigation has been demonstrated in a proof-of-concept way. For example, in October 2006, a small satellite, Earth Observing 1 (EO-1) [41], collected data on its own after noticing a plume of smoke on the island of Sumatra, Indonesia [43]. Such automatic sensor-based data collection efforts could provide insights into what happened hours before a natural hazard; in this case, before the eruption of a volcano. The overall goal is to develop objective-based rather than subjective-based models, high-resolution rather than low-resolution models, and large-scale rather than low-scale models that can form bases for extracting useful and insightful knowledge for immediate and future hazard-mitigation purposes.

This chapter is organized as follows. Section 13.2 proposes a broader knowledge discovery framework. Section 13.3 presents a brief introduction to natural hazards and sensors used for natural hazards. Section 13.4 discusses the significance and challenges of knowledge discovery from sensor data for natural hazards. Section 13.5 focuses on some applications of knowledge discovery approaches in natural hazards. Section 13.6 presents some preliminary discussions of the applications of knowledge discovery insights for hazard mitigation. Section 13.7 summarizes the chapter.

13.2 A Broader Knowledge Discovery Framework

Knowledge discovery offers tools for extracting new, useful and hidden insights from massive sensor and historic data. However, knowledge discovery techniques need to be geared towards scalable and efficient implementations of offline predictive insights, fast real-time analysis of incremental information, and solution processes for tactical and strategic decisions. Therefore, we propose a somewhat broader knowledge discovery framework (see Fig. 13.1), which describes an end-to-end process for knowledge discovery for natural disasters.

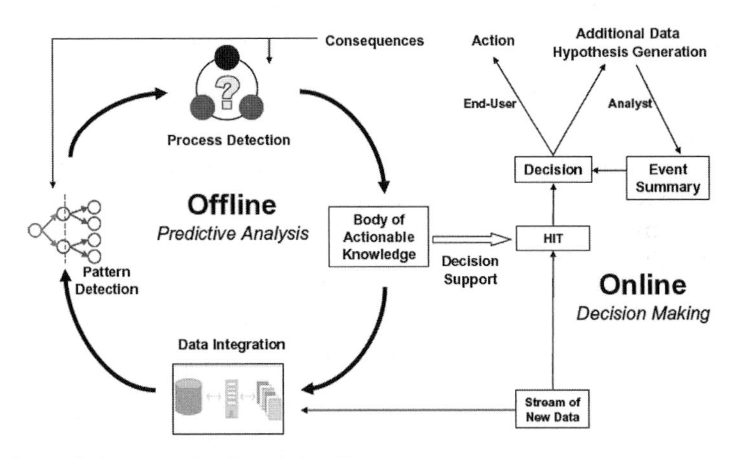

Fig. 13.1 A holistic approach to knowledge discovery

The components of the proposed framework, in the context of natural hazards, are stated in this section; a description of the two broad areas of the framework are discussed briefly in the following subsections. The proposed framework consists of two sub-frameworks:

1. Offline Predictive Analysis
   - Data Integration
     - Remote sensors, wired and wireless in-situ sensor networks

  - Numerical physics-based computer model outputs
  - Ancillary information and encoded domain knowledge
- Pattern Detection
  - Offline data mining from sensor observations and models
  - Computational efficiency and scalability to massive data
  - Anomalies, extremes, nonlinear processes, in space and time
  - Probabilities, intensities, duration, frequency, risks of observations
- Process Detection
  - Numerical models with sensor data assimilation schemes
  - Extraction of dynamics from massive sensor observations
  - Extraction of dynamics from incomplete, noisy information

2. Online Decision Making

- Decision Support
  - Online (real-time) analysis from models and observations
  - Algorithmic efficiency for dynamic, distributed processing of sensor observations
  - Resiliency, vulnerability, and impacts of observations
  - Visualization and decision or policy aids models

13.2.1 Requirements for Offline Predictive Analysis

We propose the broad requirements of the offline predictive analysis, other than the need for both capacity and capability computing, as including the following:

1. Multidisciplinary: Multiple aspects of problems solved using a set of individual tools, each motivated from one or more disciplinary area.
2. Interdisciplinary: Comprehensive solutions developed based on blend of methodologies spanning traditional disciplinary areas.
3. Process-based: Larger overall problem partitioned into component processes and solved using physics and a suite of data science tools.
4. Holistic: Approaches for an application from raw sensor and model data to decision and policy aids.

The distinguishing features compared to conventional knowledge discovery areas are the following (sub-bullets list the primary differences from the conventional):

- Data Mining
  - Enhanced focus on scientific rather than business data
  - Algorithms for anomalies, extremes, rare and unusual events rather than predicting regular events
  - Geographic, time series, spatial, space-time *relational* specific data

- Statistics and Econometrics
  - Focus on computational efficiency and scalability for distributed sensors
  - Methods for nonlinear processes and representations
  - Statistics of rare events and extremes/anomalies
- Nonlinear Dynamics and Information Theory
  - Robust to limited or incomplete and noisy information
  - Scalability to massive data for centralized sensors
  - Spatial, space-time and geographic specific data
- Signal Processing
  - Nonlinear dynamical, even chaotic, system behavior
  - Colored, even 1/f, noise
  - Noisy and incomplete information

13.2.2 Requirements for Online Decision Making

The *decision support* component is composed of online (real-time) knowledge discovery and the decision sciences. The online knowledge discovery processes need to be efficient in terms of memory usage and analysis times (especially for distributed sensors); must be able to handle incremental information in real-time; and must generate time-phased or event-based decision metrics, at multiple geographically based locations and possibly times, such that metrics can be used for automated alert mechanisms or to facilitate the task of the *human in the loop* in the space-time context.

One new example of this application is the concept of ubiquitous sensing. Ubiquitous sensing describes a situation where one has either an array of many sensors that generate high flows of data, much of which may be null (background, uninteresting or contradictory), or where one has a few mobile sensing platforms that need to be deployed in a cost-effective way. Examples of the first category include arrays that detect contraband crossing borders or unauthorized persons entering restricted areas. Examples of the latter category include satellite or air-breathing remote-sensing assets. Some offline modeling and decision-support tools have been semi-coupled in real-time to direct the next sequence of data acquisition. In one example of these decision-support tools, Bayesian approaches formulate hypotheses (such as a missile launch being detected) and marshal the data from other elements of the array (in the first example) or to move the mobile platform to the next location (in the second example) in order to gain the next most valuable data point that would reduce uncertainty once an event is detected. Some of the applications and challenges of ubiquitous sensing has been discussed in the literature [53].

Efficient real-time algorithms are required to react to the real-time, dynamic and distributed nature of knowledge discovery as well as direct this data acquisition within the time cycle of an event. Overall, the online approaches need to be algorithmically efficient, that is, the mathematical algorithms must be amenable to robust online implementations, which implies that they be fast, storage-efficient, memory efficient, adaptive and possess real-time or near-real-time analytic capacity.

However, there is a trade-off between computational efficiency, algorithm performance and domain requirements. A good example of such a trade-off is the SPIRIT algorithm [48], which is essentially an incremental version of the Principal Component Analysis (PCA) technique, but the weight estimates are slightly different from the principal directions in conventional (offline) PCA. The difference in computational requirements does not affect the algorithm performance. Therefore, a good understanding of application domains is the key to achieving such compromise. The decision-science component encompasses the development of decision metrics in space and time for visualization and visual analytics, utilization of predictive insights from offline analysis and real-time distributed discovery processes. Additionally, the decision-science component processes dynamic and event-based streams of data in conjunction with offline discovery and real-time analysis, provides feedback loops from prior decisions or policies, and provides a framework for decision metrics, uncertainty and impacts of risks, including the determination of resiliency and consequences in the context of natural disasters.

13.3 Weather, Climate, and Associated Natural Hazards

The exposure of human life and economy to natural hazards—from hurricanes, volcano, tornadoes, tsunamis and earthquakes to heat waves, cold spells, droughts and floods or flash floods—appears to have increased even as world economies have developed and prospered [50,65]. However, one natural hazard impacts the other, which generates multidimensional scenarios. In this section, we discuss impacts of some natural hazards and highlight some sensors that can collect relevant data, giving us a better understanding of the causes of, and interactions among, natural hazards.

13.3.1 Natural Hazards Impacts and Weather Extremes

Even though the occurrence of natural hazards cannot be prevented, understanding the interactions between natural hazards is significant for extracting new insights. In this section, we discuss climate impacts on weather extremes and the human impacts of weather extremes.

1. *Climate Impacts on Weather Extremes*: The tremendous uncertainty surrounding some of the issues regarding climate-weather linkage—for example, the links of global warming to the increase in the number and intensity of hurricanes—suggests that a closer inspection is necessary. Specifically, historical weather-

sensor observations and climate data gathered from various sources need to be analyzed, in an offline mode, in significant detail and with much greater care. When climate models and indicators are used to understand and quantify the impacts of climate on weather extremes, there is a need to delineate the impacts of natural climate variability before or during the quantification of the impacts of climate change. These issues are described in detail below:

- *Natural Climate Variability and Weather Extremes*: In the longer-term, natural climate variability—for example, the inter-annual El Niño phenomena—can have significant impact on weather and hydrologic extremes [19,27,57]; In fact, the 2006 hurricane season turned out to be much quieter than anticipated and the most plausible hypothesis is the occurrence of the El Niño [8], although the influence of African dust storms has also been suggested as an added factor [40]. Incidentally, as of this writing, a *very active* 2007 hurricane season is being predicted by forecasters [60].
  The need to understand and quantify the impacts of natural climate variability on weather extremes is underscored through the previous examples. The ability to quantify climate variability, including climate anomalies like El Niño, requires processing massive amounts of geographic data obtained from remote sensors like satellites and aircraft. Sensors for ocean temperature or salinity and sensor networks like ocean monitoring instrumentation also play a role. The ability to relate such large-scale geophysical phenomena to weather or hydrologic extremes and regional change, both for offline discovery and online analysis, requires holistic knowledge discovery approaches and massive computational capabilities.
  The need to quantify the impact of natural climate variability also stems from the requirement to delineate and isolate the effects of global or regional climate change [24], and in particular possible anthropogenic effects [20], on weather extremes and natural hazards.
  The impacts and current wisdom in the insurance sector (e.g., see [70] and [59], for two interesting viewpoints) may provide some indications to how the financial world may be adapting—or may be anticipating the need to adapt—to climate change. The importance of human factors rather than climate change has been emphasized as the primary driving cause for recent natural disasters [10]. Although consensus is lacking on the relative impacts of change in climate versus human factors (e.g., [59,70]), the problem points to two different lines of research. First, there is a need to understand the relative and complementary roles of climate- and human-induced changes and their combined impact on natural disaster losses. Second, there is a need to understand how future changes in climate may influence the variability of the weather extremes as well as related disaster losses. Finally, there is a need to combine the risks, consequences, vulnerabilities and anticipatory damage assessments on impacts within one policy tool which can provide metrics and visual guidance to policy makers.

- *Climate Change and Weather Extremes*: Future projections—especially at sufficiently long terms when projections based on past trends or current observations may no longer be valid—need to rely directly or indirectly on climate model simulations. State-of the-art climate models like the Community Climate System Model, Version 3, (CCSM3) can generate precise climate reconstructions and predictions. For example, CCSM3 can give estimates of climate variables at three dimensions and one-degree spatial grids, from the year 1870 to 2100 [6]. These estimates can be given at daily or even six-hour intervals However, precision does not necessarily imply accuracy, and precise predictions may be only as good as the temporal and spatial scales of the coupled atmospheric and earth system processes that such systems can model.
  The first step is to compare the model outputs with observations for time periods when both are available. This comparison is needed to understand the problems in the model outputs and to quantify the inherent uncertainties in space and time. Since any simulation model is an imperfect realization of reality and tends to smooth out the outliers and extremes, this can be a hard test for climate models. However, simplified tests may help prove a point. Thus, while the parameters of extreme value theory obtained from observations and model simulations of temperature may or may not be statistically similar, the number, frequency and duration of heat waves based on user-defined criteria and thresholds may align well and this alignment may provide sufficient information in some cases. There have been attempts to compare modeled and observed extremes [31], even based on detailed statistical analysis of extreme values [32].
  The next step is to investigate trends and patterns within climate model projection and quantify the uncertainties based on the results of model-observation comparisons. The *Science* paper by Meehl and Tebaldi [40] demonstrated how insights about future weather extremes—in their case heat waves—can be obtained in this fashion. However, this is a good starting point in terms of actionable predictive insights from a combination of observations and models. If temporally, spatially and geographically aware knowledge discovery tools [28,30,32], specifically tailored for earth science applications, are *let loose* on the massive volumes of sensor-observed and model-simulated data, we can hope to validate, and perhaps discover, insights about weather extremes. This is an urgent and high-priority research area whose time has clearly come.

2. *Human Impacts and Weather Extremes*:

   - *Globalization and Change in Human Factors*: While our discussions have focused on weather extremes alone, there have been claims that the current increase in disaster losses is more due to human factors, such as the impact of human actions on the global environment. However, there is also an understanding that anticipated climate change may begin to change the relative impacts. In any case, a link needs to be firmly established (or rejected) between the anticipated change in weather extremes, whether caused by inherent climate variability or human-induced change, to the corresponding impacts on human population [14,3] and adaptability.

- *Resiliency, Vulnerability and Policy Tools*: Weather disaster impacts relate to the design and safety of infrastructures and the resiliency of vulnerable societies [50,64,67]. An integrated policy tool is needed for various levels of strategic decisions. Thus, insights from hurricane or rainfall extremes based on archive sensor data may be used to design more resilient hydraulic structures in the short-term near the coasts, or stronger foundations for offshore structures. In addition, these policy aids can be used to assess the extreme variability, risks or consequences, resiliency and overall damage caused by anticipated extremes (for a proof-of-concept example, see [14,47,61]).
- *Policy Impacts*: A quantitative assessment of climate-weather links has direct influence on the design of highway and infrastructure sensing or monitoring systems, building redundancies for contingency planning, enhancing readiness of societies through early warning systems and public education, and planning human habitations such that vulnerabilities may be reduced. Longer-term planning based on climate projections influences human habitation and demographics through policy regulations; for example, planned movement of populations from vulnerable regions. This may be a feedback loop as the quantitative assessment may help guide climate policy.

13.3.2 Utilization of Sensors for Weather and Climate

The occurrence of natural hazards cannot be prevented; but their occurrences and interactions can be studied for useful insights into what drives them. Sensors have long been used to collect data about weather (e.g., DOPPLER), climate, and natural hazards. Recent advances in computational techniques and the current advances in satellite, telecommunication and sensor technologies are providing access to, and analysis of, massive data that can provide better knowledge of what drives these hazards. The types of natural hazards are many but the most common (in alphabetical order) include asteroid, avalanche, drought, earthquake, flood, heat wave, hurricane, landslide, salinity, tornado, tsunami, volcanism, and wildfire. Each of these natural hazards has been an integral part of the human experience. However, their occurrences, as well as their effects on lives, properties and infrastructures, are becoming more dramatic. For example, drought may not be the most dramatic occurrence, but it is one of the most damaging disasters. Since 1967, drought alone has been responsible for millions of deaths and has cost hundreds of billions of dollars in damage worldwide [41]. Some of the sensors that can be used to track these hazards and collect related data are:

1. DOPPLER Radars: These are weather-related sensors that send out radio waves and are one of the oldest weather-related sensors.
2. Earth Observing Sensors: These are satellite-based sensors used by NASA to monitor the Earth. An example of these sensors is Earth Observing 1 (EO-1).

These sensors are useful for weather/climate-related hazards such as heat waves, volcano, and hurricanes [41].

3. The National Ecological Observatory Network (NEON) sensors: This array of sensors are used to understand how land-use change and climate variation affect ecological systems [22].
4. Remote Sensors: Remote sensors are used to measure global ice cover changes and carbon deposits, which can help track hurricanes, forest fires, and many other climate-related hazards. Two types of remote sensing are used by NASA: passive remote sensing, such as radiometers, and active remote sensing, such as RADAR and Lidar [42].
5. River Sensor Network: This network of sensors developed at the University of Lancaster monitors water depth and flow, which can be used to predict impending floods. Some of the sensors in the network measure pressure from below the water line in order to determine depth; others monitor the speed of river flow to track objects and ripples moving along the surface from the riverbank [44].
6. Satellite Imagery: Satellite imagery is a multi-sensor system useful for capturing forces of nature such as hurricanes [11].

Other examples are in-situ sensors, such as the Prompt Assessment of Global Earthquakes for Response (PAGER) system, developed by the US Geological Survey, which automatically estimates human impact following significant earthquakes [66]. This system also provides important information to help emergency relief organizations, government agencies and the media plan their response to earthquake disasters. There are also efforts to distribute data from environmental and ecological sensors to interested communities for further analysis. For example, the sea-viewing Wide Field-of-view Sensor (SeaWiFS) project provides quantitative data on global ocean bio-optical properties to the earth science community [62]. These data are useful for extracting insightful knowledge that can help us understand the causes driving these hazards. These sensors and others provide data that can be used for insightful predictions of when an hazard will happen and the potentially affected areas.

13.4 Challenges of Knowledge Discovery from Sensor Data for Natural Hazards

The earth science community has developed and used traditional statistics [71], non-traditional statistical models like extreme value theory [25,26], spatial and space-time statistics [7,45], and nonlinear dynamics and signal processing [46,54,71], for conventional data-analysis applications. However, the availability of advanced sensor technology with high-performance computational facilities provides opportunities for developing solutions beyond statistical analysis of isolated data sets. Some applications of knowledge discovery for the earth sciences have been reported in the last few years [21,52,63]. However, in view of the amount of sensor data available, contributions from data mining have been limited. For example, the process of using remotely sensed and normalized difference vegetation index (NDVI) data for land cover change detection is well known. Potere et al. [51] demonstrated a novel knowledge discovery approach where time series data for NDVI (derived from the MODIS sensor [37] between 2000 and 2005) was used to detect changing landscape and land use from construction of WalMart stores. However, such an approach, originating from traditional earth science perspective, was rather visually driven and not easily transferable from a knowledge discovery *approach* to a knowledge discovery *process* through online implementation of statistical reasoning. As a result, an online knowledge discovery approach was developed by Fang et al. [12] as a first step towards automating landscape and land-use change detection process. The online approach was motivated by statistical process control methods for change detection. The automated approach, which is an adaptation of simulated annealing for change-point detection, was validated with WalMart store openings data (Fig. 13.2) and has encouraging results.

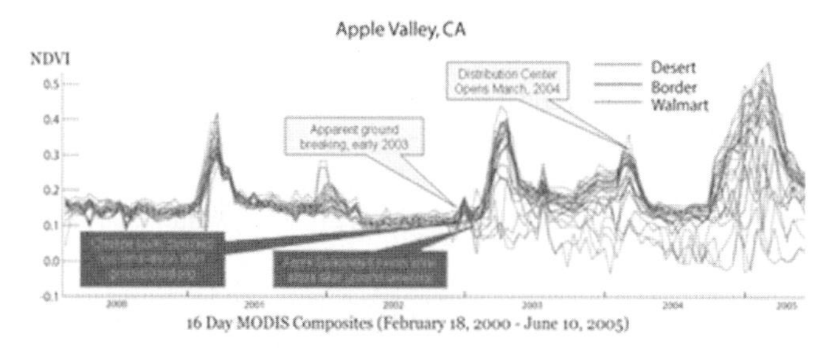

Fig. 13.2 Online change detection and alarms [12]

The online approach is a real-time approach in the sense that incremental data can be analyzed efficiently as soon as they become available. An extension of this approach can be used in the context of real-time natural disaster management by identifying regions in space and time with significant and rapid change in land cover. The analysis methods can be used to investigate geographic and remote sensing data and then zero in on areas where change is occurring or has occurred in the recent past. In addition, the methods can be used to identify changes due to deforestation. The efficiency of the approach and its incremental approach differentiates it from many traditional approaches used in earth science and remote sensing change detection applications. The interdisciplinary online approach [12] is efficient and lends itself to full automation unlike visualization-based manual validation [51] traditionally used in remote sensing. In addition, the ability to make immediate decisions based on incremental information is a distinguishing feature. The approach can be further developed to automatically detect land cover changes from large-scale and high-resolution geospatial-temporal data, as well as to automatically zoom in on the specific locations where such change may have occurred. This capability is important to assess natural disaster damage by investigating remotely sensed images before and after an event.

Newer developments in data mining include the ability to deal with dependence of learning samples or "relational" data [9,39,55], mining rare events from dynamic sequences [68,69], as well as anomaly detection (primarily in the context of cyber-security, e.g., [34]) and change-point detection [18].

Computationally efficient methods for anomalies, extreme values, rare events, change and nonlinear processes need to be developed for massive geographically based sensor data. An interdisciplinary focus within the data sciences is necessary, so that new tools can be motivated from a combination of traditional and non-traditional statistics, nonlinear dynamics, information theory, signal processing, econometrics and decision theory, and lead toward application solutions that span these areas. One key challenge of natural hazards is the multi-scale nature of the problem, which may mean that the physical processes, parameterization and parameter values, as well as the fitted data-generation models or distributions, do not remain invariant across space-time scales. In a section describing the *interdisciplinary nature of knowledge discovery in databases (KDD)*, Fayyad et al. [13] mentioned that KDD evolves from the *intersection of research fields* like *machine learning, pattern recognition, databases, statistics, AI, knowledge acquisition for expert systems, data visualization, and high performance computing.*

The time may have come to further broaden the interdisciplinary roots. Scientific applications are built on domain knowledge and are typically embedded with numerical models. The sensor technology and knowledge discovery communities also have to work together with modelers to develop efficient strategies for real-time data assimilation within physically based computer models. In addition, analysis from observations needs to be effectively synthesized with model simulations. While the challenges are tremendous, the scientific opportunities [4] and benefits can be immense.

Some of the reasons there have been limited research efforts in weather and climate extremes using knowledge discovery approaches are:

1. *Limited multidisciplinary span*: There is a need to extend knowledge discovery approaches to include tools from other disciplinary areas such as spatial statistics, extreme value theory and nonlinear dynamics, information theory and decision sciences.
2. *Limited Interdisciplinary span*: Solutions should be based on a blend of methodologies including traditional areas of statistics and machine learning.
3. *Lack of holistic-based solutions*: The knowledge discovery approaches should be formulated with solutions in mind rather isolated predictive analyses.
4. *Approaches for strategic and tactical decisions*: The knowledge discovery approaches must be focused on both short-term decisions and long-term planning rather than immediate implications.

As a result, we propose a broader and holistic knowledge discovery framework in Sect. 13.2.

13.5 Knowledge Discovery Approaches for Weather, Climate and Associated Natural Hazards

This section provides an overview of the knowledge discovery approaches developed or being developed by the authors and their collaborators in the area of weather and climate extremes and related natural hazards. While the sample approaches are neither exhaustive nor intended to be directional, they illustrate the areas where knowledge discovery, broadly construed, can help in the context of natural hazards.

The examples presented in this section attempt to emphasize the potential of knowledge discovery approaches. In addition, these are intended to provide examples of both a closed-loop knowledge discovery process depicted in Fig. 13.1 as well as for knowledge discovery approaches that may be useful for decision and policy making in natural hazards mitigation.

13.5.1 Knowledge Discovery Approaches for Weather, Climate and Associated Natural Hazards

This section provides an overview of the knowledge discovery approaches developed or being developed by the authors and their collaborators in the area of weather and climate extremes and related natural hazards. While the sample approaches are neither exhaustive nor intended to be directional, they can help illustrate the possibilities that knowledge discovery, broadly construed, can help in the area of natural hazards.

The examples presented in this section attempt to emphasize the potential of knowledge discovery approaches. In addition, these are intended to provide examples of both a closed-loop knowledge discovery process depicted in Figure 1.1 as well as for knowledge discovery approaches that may be useful for decision and policy making in natural hazards mitigation.

Some of the building blocks which can ultimately lead to a closed-loop process for tactical and strategic decision-making in the context of weather or climate related hazards are:

1. Short-term prediction from remotely sensed observations [16].
2. Trends in weather extremes from dynamic data streams [30].
3. Prediction from short and noisy sensor data [29].
4. Natural variability and impacts on local geophysical phenomena [27].
5. Comparison of model simulations and historical observations [32].
6. Real-time change detection from remotely sensed data [12].
7. Quantification and visualization of human impacts [14].

Here, we highlight a few of these building blocks through the following results:

1. *Short-term prediction from remotely sensed observations*: Figure 13.3 depicts the short-term rainfall prediction methodology developed by Ganguly and Bras [16].

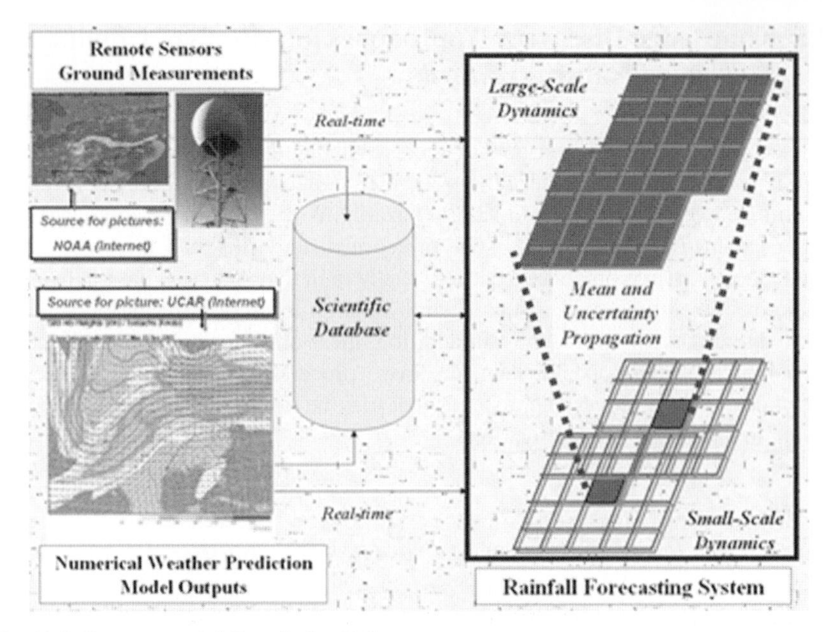

Fig. 13.3 Short-term rainfall prediction [16]

Numerical weather prediction model outputs and remote sensor observations from weather radar were blended for high-resolution forecasts at radar resolutions and for one- to six-hour lead times.
The approach relied on a process-based strategy, where the overall problem was partitioned into component processes based on domain knowledge and exploratory data analysis and then the results were re-combined. The forecasting strategy used a combination of weather physics like advection, as well as a suite of traditional and new adaptations of data-dictated tools like exponential smoothing, space-time disaggregation and Bayesian neural networks. Case studies with real data indicated [15,16] that the methodology was able to outperform the state-of-the-art approach at that time. The ability to generate short-term (0–6 hour) and high-resolution (order of a km or less in space and hourly or less in time) quantitative precipitation forecasts, especially for convective storms, is important for heavy rainfall events and hurricane activity, primarily to quantify the potential risks and damage from flash flood and flood-related hazards. Advance information can be used to control hydraulic flows, take preparatory measures and issue flood advisories.

2. *Trends in weather extremes from dynamic data streams*: Figure 13.4 exhibits a preliminary result from the approach developed by [30] for computing the spatio-temporal trends in the volatility of precipitation extremes. The methods were used for large-scale, geographically dimensioned data at high spatial resolutions. The geospatial-temporal indices were computed at each grid point in South America for which rainfall time series was available. The change in the indices can be

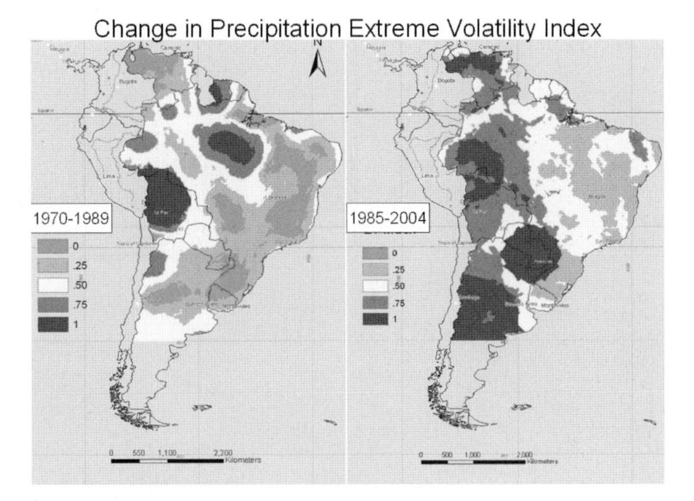

Fig. 13.4 Space-time trends in extremes volatility [30]

quantified and visualized with multiple time windows of data. The color scheme in Fig. 13.4 and the GIS-based visualization was done as part of the [14] study described later.

The red-amber-green color combination is used to denote high (red) to low (green) volatility for the extremes. The extremes volatility index was a new measure based on the ratios of return levels, computed at each grid. The index presented here was normalized to scale between zero and unity by [14]. The ability to quantify and visualize weather and hydrologic extreme values and their properties (e.g., 100-year levels) in space and time is an important first step to studying the impacts on these extremes on infrastructures and human societies. One implication of this study for natural hazards is that it can help evaluate the threat posed by failure by critical infrastructures, such as dams. The extreme volatility index is a measure of the anticipated degree of surprise, or "threat", due to natural extremes. The measure relates, in an aggregate sense, to the expected impacts of extremes. Thus, if critical infrastructures such as dams or levees have been designed to withstand rare 100-year rainfall events (or a rainfall intensity of 0.01 probability of exceedance) then a rarer and more intense event (e.g., a 500-year rainfall) may cause significant damage only if the 500-year intensity is significantly different from the 100-year intensity. This second-order information about relative intensity of extremes is important both for natural variability of the climate system and in situations where global change may cause the extremes to grow more intense. The information can be used for risk-benefit analysis during the design of hydraulic structures and response systems.

3. *Prediction from short and noisy sensor data*: The ability to deal with massive volumes of geographic data from remote and in-situ sensors needs to be complemented by the ability to derive predictive insights from short and noisy geophysical and weather- or climate-related observations. Some of the most relevant his-

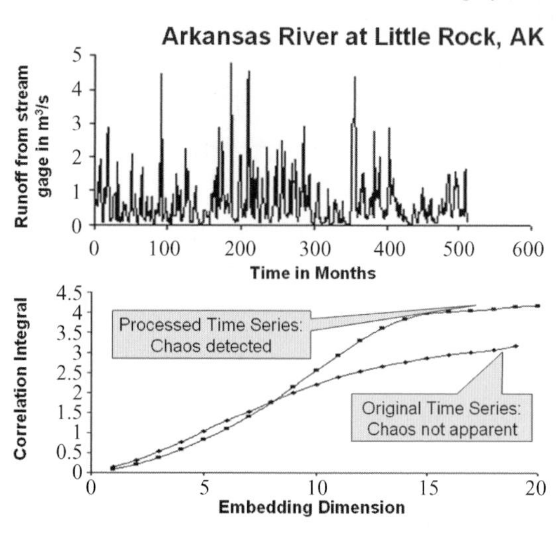

Fig. 13.5 Predictability in short and noisy data [29]

torical geophysical data and indices may be limited or incomplete, however the presence of nonlinear dynamics and chaos on the one hand, and colored or even 1/f noise with seasonal fluctuations on the other, cannot be ruled out a priori. In fact, the ability to detect the underlying nonlinear dynamical signals from such data may be of significant value for studies in short- and long-term predictability. Khan et al. [29] developed a methodology to extract the seasonal, random and dynamical components from short and noisy time series, and applied the methods to simulated data and real river flows. Figure 13.5 shows the application to the Arkansas River. The methodology was based on a combination of tools from signal processing, traditional statistics, nonlinear prediction and chaos detection. This methodology can be used to determine how much information is actually contained in the data; such quantification can help determine the appropriate preprocessing approaches and which knowledge discovery techniques to use. The ability to quantify how much predictive insights can be generated from data has direct implications for anticipatory risk-mitigation strategies. Thus, when the available data are completely random, a risk-benefit analysis based on standard deviations may be appropriate, while for completely deterministic signals, an investment in the development of better predictive models followed by recommendations of specific mitigation strategies may be the better strategy. However, for nonlinear dynamics and chaos, the trade-offs between short-term predictability and longer-term error growth need to be carefully balanced, depending on how much information can be extracted from data, especially when the data are noisy and/or limited. Thus, information on the type, quality and quantity of predictive insights that can be generated from data may lead to a determination of preventive actions that can be taken in anticipation of climate, weather and hydrologic extremes.

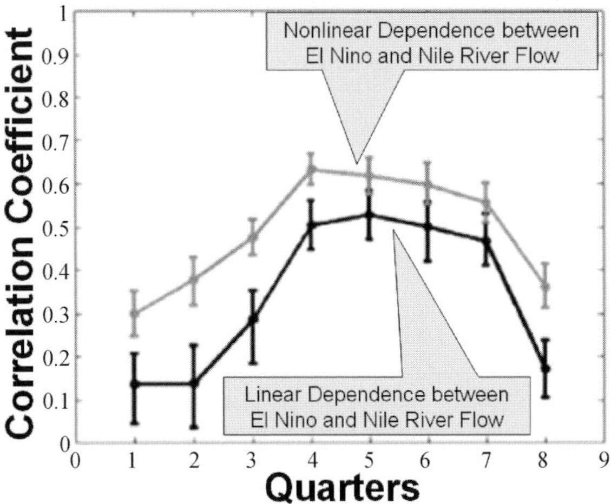

Fig. 13.6 Nonlinear impact of El Niño on rivers [27]

4. *Natural variability and impacts on local geophysical phenomena*: The ability to quantify nonlinear dependence from historical data, even in situations where such data are short and noisy, are critical first steps in studies on predictability, predictive modeling, and physical understanding of weather, climate and geophysical systems.
   Khan et al. [27] developed an approach based on nonlinear dynamics and information theory, along with traditional statistics, to develop and validate new adaptations of emerging techniques. The approach was tested on simulated data, and then applied to an index of the El Niño climate phenomena and the variability in the flow of tropical rivers. The approach revealed more dependence between the variables than previously thought. The ability to quantify the impacts of natural climate variability on weather and hydrologic variables, as shown in Fig. 13.6, can help refine our understanding of the impacts of climate change. The methodology, which can have significant broader impact beyond the case study considered here, was refined and expanded in a another work [28]. Climate systems' response to global changes often leads to natural hazards. We note that the individual and combined impacts of El Niño and global warming have often been advanced as causes for relatively hot or cold summers, as well as the activity of the hurricanes season, in the continental United States. An extraction of causality may be an open research area; however, previous researchers have suggested that natural variability in climate systems, as well as global environmental change, may cause hydrologic or weather extremes at local or regional scales. The ability to quantify the dependency among natural or changing climate phenomena and natural extremes or hazards can help point to appropriate information sources that may guide predictive analyses. This is especially true when larger-scale climate effects can be predicted in advance from data or from

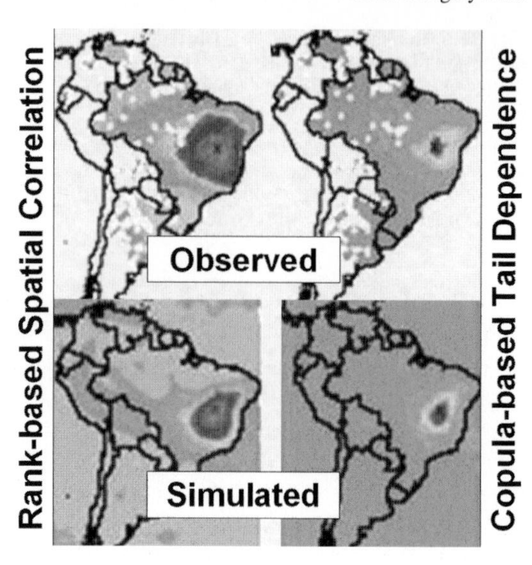

Fig. 13.7 Geospatial-temporal extreme dependence [32]

simulations, which in turn can be used to provide predictive insights on natural hazards.

5. *Comparison of model simulations and historical observations*: Kuhn et al. [32] developed a new approach to quantify the geospatial-temporal dependence among extreme values from massive geographic data. The methodology was motivated by recent developments in multivariate extremes, and hence can be applied to quantify the dependence among extremes of multiple variables—for example, heat waves and precipitation extremes—in space and time. In addition, this copula-based measure can be useful in analyzing simultaneous occurrence of extremes, which may be indicators of possible change.
Thus, if two 100-year events which have zero extremes dependence were to occur simultaneously, this would be a 10,000-year event (see [32], for details), whereas if the extremes have complete dependence, then the simultaneous occurrence still represents a 100-year event. The new measure was utilized on geospatial-temporal rainfall observations and climate model simulations for intercomparisons and model evaluation as shown in Fig. 13.7. In addition, the extremes dependence in space and time was compared with the corresponding spatial correlation values, obtained here through a rank-based correlation measure. Co-occurrence of extremes like heat waves and prolonged droughts can have a combined impact on human lives and economies that is greater than the sum of the individual impacts. The co-occurrence of extremes over space and time may imply a larger regional impact, for example, co-occurrence of extreme rainfall over larger areas may increase the chances of widespread flooding. The relation among extremes in time can be useful for predictive insights regarding the extremes of one variable based on observations of extremes in related variables. The simultaneous and/or frequent occurrence of multiple extremes in space and

time may suggest local, regional or global change in the underlying dynamics of the weather or climate system.

13.6 The Significance of Utilizing Knowledge Discovery Insights for Hazards Mitigation

While debates may persist on the exact causes of natural hazards that led to increased losses of human life and property in recent years, the fact that enhanced predictive insights can help mitigate the impacts of such hazards through improved decisions and policies is becoming relatively well accepted. Pearce [49] described the need for a *shift [in] focus from response and recovery to sustainable hazard mitigation* and laments that in current practices *hazard awareness is absent from local decision-making processes*. However, such sustainable hazard mitigation depends on useful predictive insights from historical data. An interesting early article by Sarewitz and Pielke, Jr. [58] discusses the art of making scientific predictions relevant to policy makers, and uses weather extremes and natural hazards as examples of possible applications. The idea is to move beyond post-disaster consequence management and humanitarian aid disbursal toward preemptive policies based on predictive insights. Decision makers need to realize that while natural disasters may or may not be *acts of God*, their consequences affect humans and can be mitigated by policies; this was highlighted by two recent *Science* magazine articles [3,36].

Hazards mitigation, using predictive insights in some rudimentary form, has been attempted since time immemorial with varying success. However, what has changed dramatically in recent years is the availability of massive volumes of historical and real-time data from sensors. These data, combined with advanced techniques and high-performance computational tools that can be used to extract actionable predictive insights, enhance our understanding of physical processes. This leads to improved short- and longer-term computer simulation models, which in turn are initialized and updated in real time with sensor observations for improved accuracy. The result is a yield of massive volumes of simulation outputs. In this sense, both the new opportunities and the key challenges in generating predictive insights for natural hazard mitigation rely on extracting knowledge from massive volumes of dynamic and distributed sensory data as well as large volumes of computer-based simulations.

An initial investigation in this area was performed by Fuller et al. [14]. Their work investigated how a combination of variables—specifically, the precipitation extremes volatility as defined by Khan et al. [30], the high-resolution population maps described by Bhaduri et al. [2] and used by Sabesan et al. [56], as well as measures representing development or financial indices like the GDP—could be used in conjunction with each other to quantify and visualize the human impacts on natural disasters, specifically those caused by rainfall extremes in South America. Figure 13.8 is a map showing the impacts of weather-related disaster on the human population based on their investigation.

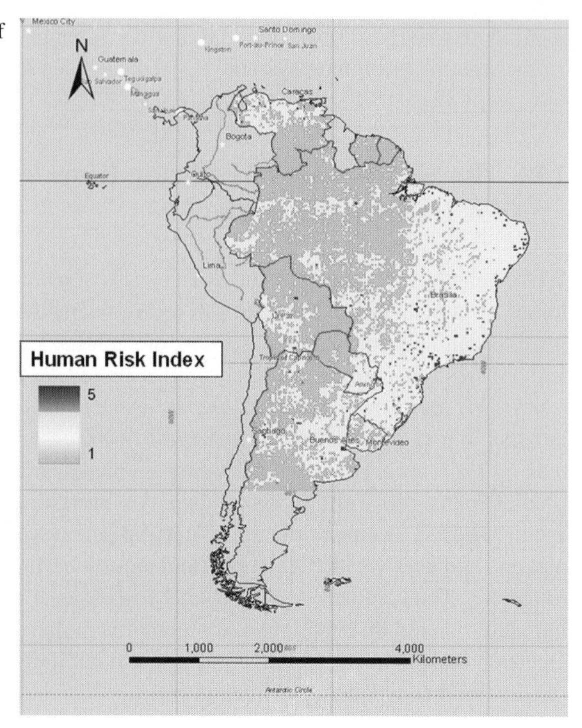

Fig. 13.8 Human impacts of weather related disasters [14]

The computed geospatial indices included the probabilities of truly unusual rainfall extremes, the risks to human population associated with such extremes and the resiliency, or the ability of a region to respond to the disaster. Anticipatory information on disaster damage based on refinements of this study can aid policy makers. Risk metrics can be designed and quantified in space and time based on *threat* or degree of surprise caused by natural disasters, as well as consequences to human population, economies and critical infrastructures. Resiliency metrics for infrastructures and societies can be used in conjunction with risks to develop geospatial and temporal metrics for anticipated impacts. The metrics can provide an overall and objective assessment of potential disaster damage to emergency planners and policy makers at high-resolutions in space and time over large space-time scales. In addition, the various metrics can help planners perform root-cause analysis to determine the critical responsible factors. The overall assessment can help policy makers optimize the level of resource allocations in space and time while the root-cause analysis can help design appropriate mitigation strategies based on the allocated resource at any specific location in any given time.

13.7 Closing Remarks

Predictive insights generated from sensor data, in conjunction with data obtained from other sources like computer-based simulations, can facilitate short-term decisions and longer-term policies. The overall process where raw data from sensors or simulations are ultimately converted to actionable predictive insights for decision and policy makers is defined as knowledge discovery in this chapter. This chapter presents a broader view of knowledge discovery compared to the traditional definitions by the data mining community, but with the data mining and other data sciences as key aspects of the overall process. In addition, we have defined sensors broadly to include wireless sensor networks, in-situ sensor infrastructures and remote sensors. The challenges and opportunities for knowledge discovery based on data from sensors and simulations were described. In particular, we have presented a vision of knowledge discovery in the context of scientific applications. This chapter describes how knowledge discovery from historical and real-time sensor data and computer model simulations can lead to improved predictive insights about weather, climate and associated natural hazards, which can in turn be combined with metrics for disaster risks, consequence, and vulnerability. Scientific applications and scientific knowledge discovery may make sense primarily in the context of a specific domain. Our focus is weather, climate and geophysical hazards. While prediction of natural hazards and mitigating their consequences have been attempted since the dawn of human civilization with varying degrees of success, the possibility of enhanced knowledge discovery from ever-increasing and improving sensor and simulation data make us optimistic that significant breakthroughs may be possible in the near future.

Acknowledgements This research was funded by the Laboratory Directed Research and Development (LDRD) Program of the Oak Ridge National Laboratory (ORNL), managed by UT-Battelle, LLC, for the U.S. Department of Energy under Contract DE-AC05-00OR22725. We are grateful to Dr. Vladimir Protopopescu, Dr. David J. Erickson III and Mark Tuttle, all of ORNL, for their reviews of the manuscript within ORNL's internal publication tracking system. In addition, we are thankful to Drs. Ranga Raju Vatsavai and Steven J. Fernandez, both of ORNL, for their informal reviews and comments. We would also like to thank our co-authors and others who have helped and are continuing to help in our recent and ongoing research highlighted in Sect. 13.6 of this chapter. In particular, we thank Dr. Gabriel Kuhn, Aarthy Sabesan, Kathleen Abercrombie, and Christopher Fuller, who worked on some of the cited projects while Gabriel and Aarthy were at ORNL as post-masters associates, and Kate and Chris as undergraduate interns, respectively. We are grateful to Drs. Vladimir Protopopescu and Brian Worley of ORNL for their contributions to Fig. 13.1, to Professor George Cybenko of Dartmouth for the process detection emblem, and David Gerdes, formerly of ORNL, for his help with the figure. We also gratefully acknowledge the suggestions by the original *PRIDE* team comprising collaborators at ORNL and multiple university partners for their helpful suggestions. Table 1 is an adapted version of the requirements proposed by the first author at a working group meeting of the Weather Extremes Impacts Workshop organized by Los Alamos National Laboratory and the National Center for Atmospheric Research at Sante Fe on February 27–28, 2007. Shiraj Khan and Auroop Ganguly would like to acknowledge the help and support of Professor Sunil Saigal, currently at the University of South Florida (USF), for his support, and for facilitating the completion of Shiraj Ph.D. program at USF with Auroop as his major supervisor. This chapter has been authored by UT-Battelle, LLC, under contract DE-

AC05-00OR22725 with the U.S. Department of Energy. The United States Government retains and the publisher, by accepting the article for publication, acknowledges that the United States Government retains a non-exclusive, paid-up, irrevocable, world-wide license to publish or reproduce the published form of this manuscript, or allow others to do so, for United States Government purposes.

References

[1] A. Agovic, A. Banerjee, A.R. Ganguly, V.A. Protopopescu, Anomaly detection in transportation corridors using manifold embedding. In: Proceedings of the First International Workshop on Knowledge Discovery from Sensor Data, ACM KDD Conference, San Jose, CA.
[2] B. Bhaduri, E. Bright, P. Coleman, J. Dobson, LandScan: locating people is what matters. Geoinformatics, 5(2):34–37, 2002.
[3] J. Bohannon, Disasters: searching for lessons from a bad year. Science, 310(5756):1883, 2005.
[4] C.E. Brodley, L. Terran, T.M. Stough, Knowledge discovery and data mining, computers taught to discern patterns, detect anomalies and apply decision algorithms can help secure computer systems and find volcanoes on Venus. American Scientist, 87(1):54, 1999.
[5] D. Butler, 2020 computing: everything, everywhere. Nature, 440:402–405, 2006. doi: 10.1038/440402a.
[6] W.D. Collins, C.M. Bitz, M.L. Blackmon, G.B. Bonan, C.S. Bretherton et al., The community climate system model version 3 (CCSM3). Journal of Climate, 19(11):2122–2143, 2006.
[7] N. Cressie, Statistics for Spatial Data. Wiley–Interscience, New York, 1993.
[8] Drye, Mild U.S. hurricane season defied predictions. National Geographic, 30 November 2006.
[9] S. Dzeroski, N. Lavrac, Relational Data Mining. Springer, Berlin, 2001.
[10] D.R. Easterling, G.A. Meehl, C. Parmesan, S.A. Changnon, T.R. Karl, L.O. Mearns, Climate extremes: observations, modeling and impacts. Science, 289(5487):2068–2074, 2000.
[11] ESA http://www.esa.int/esaCP/SEMPMB0XDYD_index_0.html. Downloaded: 30 May 2007.
[12] Y. Fang, A.R. Ganguly, N. Singh, V. Vijayaraj, N. Feierabend, D.T. Potere, Online change detection: monitoring land cover from remotely sensed data, ICDMW, In: Sixth IEEE International Conference on Data Mining—Workshops (ICDMW'06), pp. 626–631, 2006.
[13] U. Fayyad, G. Patietsky-Shapiro, P. Smyth, From data mining to knowledge discovery in databases. AI Magazine, 17(3):37–54, 1996.
[14] C. Fuller, A. Sabesan, S. Khan, A. Kuhn, G. Ganguly, D. Erickson, G. Ostrouchov, Quantification and visualization of the human impacts of anticipated precipitation extremes in South America, Eos Transactions, American Geophysical Union, 87(52), Fall Meeting Supplement, Abstract GC44A-03, 2006.
[15] A.R. Ganguly, A hybrid approach to improving rainfall forecasts. Computing in Science and Engineering, 4(4):14–21, 2002.
[16] A.R. Ganguly, R.L. Bras, Distributed quantitative precipitation forecasting using information from radar and numerical weather prediction models. Journal of Hydrometeorology, 4(6):1168–1180, 2003.
[17] R.L. Grossman, C. Kamath, P. Kegelmeyer, V. Kumar, R.R. Namburu, Data Mining for Scientific and Engineering Applications. Kluwer Academic, Dordrecht, 2001.
[18] V. Guralnik, J. Srivastava, Event detection from time series data. In: Proceedings of the Fifth ACM SIGKDD International Conference on Knowledge Discovery and Data Mining, pp. 33–42, 1999.
[19] A. Hannachi, Quantifying changes and their uncertainties in probability distribution of climate variables using robust statistics. Climate Dynamics, 27(2–3):301–317, 2006.

[20] G. Hegerl, C. Zwiers, F.W. Stott, V.V. Kharin, Detectability of anthropogenic changes in annual temperature and precipitation extremes. Journal of Climate, 17(19):3683–3700, 2004.
[21] T.H. Hinke, J. Rushing, H. Ranganath, S.J. Graves, Techniques and experience in mining remotely sensed satellite data. Artificial Intelligence Review, 14(6):503–531, 2000.
[22] M. Hopkin, Ecology: Spying on nature. Nature, 444:420–421, 2006.
[23] C. Huang, T. Hsing, N. Cressie, A.R. Ganguly, V.A. Protopopescu, N.S. Rao, Statistical analysis of plume model identification based on sensor network measurements. Accepted by ACM Transactions on Sensor Networks, to be published, 2007.
[24] IPCC, Climate Change 2007: The Physical Science Basis. Summary for Policymakers, Intergovernmental Panel on Climate Change, 18 pages, 2007.
[25] R.W. Katz, Stochastic modeling of hurricane damage. Journal of Applied Meteorology, 41:754–762, 2002.
[26] R.W. Katz, M.B. Parlange, P. Naveau, Statistics of extremes in hydrology. Advances in Water Resources, 25:1287–1304, 2002.
[27] S. Khan, A.R. Ganguly, S. Bandyopadhyay, S. Saigal, D.J. Erickson, III, V. Protopopescu, G. Ostrouchov, Nonlinear statistics reveals stronger ties between ENSO and the tropical hydrological cycle. Geophysical Research Letters, 33:L24402, 2006.
[28] S. Khan, S. Bandyopadhyay, A.R. Ganguly, S. Saigal, D.J. Erickson, III, V. Protopopescu, G. Ostrouchov, Relative performance of mutual information estimation methods for quantifying the dependence among short and noisy data. Physical Review E, to appear, 2007.
[29] S. Khan, A.R. Ganguly, S. Saigal, Detection and predictive modeling of chaos in finite hydrological time series. Nonlinear Processes in Geophysics, 12:41–53, 2005.
[30] S. Khan, G. Kuhn, A.R. Ganguly, D.J. Erickson, G. Ostrouchov, Spatio-temporal variability of daily and weekly precipitation extremes in South America. Water Resources Research, to appear, 2007.
[31] D. Kiktev, D.M.H. Sexton, L. Alexander, C.K. Folland, Comparison of modeled and observed trends in indices of daily climate extremes. Journal of Climate, 16(22):3560–3571, 2003.
[32] G. Kuhn, S. Khan, A.R. Ganguly, M. Branstetter, Geospatial-temporal dependence among weekly precipitation extremes with applications to observations and climate model simulations in South America. Water Resources Research, to appear, 2007.
[33] K.E. Kunkel, R.A. Pielke Jr., S.A. Changnon, Temporal fluctuations in weather and climate extremes that cause economic and human health impacts: a review. Bulletin of the American Meteorological Society, 80:1077–1098, 1999.
[34] W. Lee, S.J. Stolfo, K.W. Mok, A data mining framework for building intrusion detection models. IEEE Symposium on Security and Privacy, sp, p.0120, 1999.
[35] T.M. Lillesand, R.W. Kiefer, J.W. Chipman, Remote sensing and image interpretation, 5th edn. Wiley, New York, 2004.
[36] J. Linnerooth-Bayer, R. Mechler, G. Pflug, Refocusing disaster aid. Science, 309(5737):1044–1046, 2005.
[37] MODIS, Moderate Resolution Imaging Spectroradiometer. http://modis.gsfc.nasa.gov/about/, 2007.
[38] J. McEnery, J. Ingram, Q. Duan, T. Adams., L. Anderson, NOAA's advanced hydrologic prediction service: building pathways for better science in water forecasting. Bulletin of the American Meteorological Society, 86:375–385, 2005.
[39] H.J. Miller, J. Han, Geographic Data Mining and Knowledge Discovery. Taylor and Francis, London, 2001.
[40] NASA, Did dust bite the 2006 hurricane season forecast? NASA News, 2007.
[41] NASA, Earth Observatory Natural Hazards. http://earthobservatory.nasa.gov/NaturalHazards. Downloaded: 30 May 2007.
[42] NASA, Earth Science Remote Sensing. http://science.hq.nasa.gov/earth-sun/technology/remote_sensing.html. Downloaded: 30 May 2007.
[43] NASA, A Growing Intelligence Around Earth, Science@NASA, Headline News Feature, 2006.

[44] NewScientistTech, http://www.newscientisttech.com/article/dn10360. Downloaded: 30 May 2007.
[45] D. Nychka, C.K. Wikle, J.A. Royle, Multiresolution models for nonstationary spatial covariance functions. Statistical Modelling: An International Journal, 2:315–331, 2002.
[46] G.R. North, K.-Y. Kim, S.S.P. Shen, J.W. Hardin, Detection of forced climate signals. Part I: Filter theory. Journal of Climate, 8(3):401–408, 1995.
[47] Oak Ridger, ORNL project aims to lessen impact of natural disasters, Oak Ridger, 2007.
[48] S. Papadimitriou, J. Sun, C. Faloutsos, Streaming pattern discovery in multiple time-series, VLDB, pp. 697–708, 2005.
[49] L. Pearce, Disaster management and community planning, and public participation: how to achieve sustainable hazard mitigation. Natural Hazards, 28(2–3):211–228, 2003.
[50] M. Pelling, Natural disasters and development in a globalizing world. Rutledge, 2003.
[51] D. Potere, N. Feierabend, E. Bright, A. Strahler, Walmart from space: a new source for land cover change validation. Photogrametric Engineering and Remote Sensing, accepted for publication, 2007.
[52] C. Potter, P.-N. Tan, M. Steinbach, S. Klooster, V. Kumar, R. Myneni, V. Genovese, Major disturbance events in terrestrial ecosystems detected using global satellite data sets. Global Change Biology, 9(7):1005–1021, 2003.
[53] D. Puccinelli, M. Haenggi, Wireless sensor networks: applications and challenges of ubiquitous sensing. IEEE Circuits and Systems Magazine, 5(3):19–31, 2005.
[54] R.H. Reichle, D.B. McLaughlin, D. Entekhabi, Hydrologic data assimilation with the Ensemble Kalman filter. Monthly Weather Review, 130(1):103–114, 2002.
[55] J.F. Roddick, K. Hornsby, M. Spiliopoulou, An updated bibliography of temporal, spatial and spatio-temporal data mining research. Lecture Notes in Computer Science, 147 pages, 2001.
[56] A. Sabesan, K. Abercrombie, A.R. Ganguly, B.L. Bhaduri, E.A. Bright, P. Coleman, Metrics for the comparative analysis of geospatial datasets with applications to high resolution grid-based population data. GeoJournal, to appear, 2007.
[57] A. Sankarasubramanian, U. Lall, Flood quantiles in a changing climate: seasonal forecasts and causal relations. Water Resources Research, 39(5):1134, 2003.
[58] D. Sarewitz, R. Pielke, Jr., Prediction in science and policy. Technology in Society, 21:121–133, 1999.
[59] Q. Schiermeier, Insurers' disaster files suggest climate is culprit: rising costs hint at weather effect. Nature, 441:674–675, 2006.
[60] ScienceDaily, Very active 2007 hurricane season predicted. Science Daily, 2007.
[61] ScienceDaily, Study aims to ease natural disaster impact. Science Daily, 2006.
[62] SeaWiFS, Background of the SeaWiFS Project. http://oceancolor.gsfc.nasa.gov/SeaWiFS. Downloaded, 30 May 2007.
[63] M. Steinbach, P.-N. Tan, V. Kumar, C. Potter, S. Klooster, Discovery of climate indices using clustering. In: Proceedings of the Ninth ACM SIGKDD International Conference on Knowledge Discovery and Data Mining, pp. 446–455, 2003.
[64] K.J. Tierney, M.K. Lindell, R.W. Perry, Facing the Unexpected: Disaster Preparedness and Response in the United States. The National Academies Press, 318 pages, 2001.
[65] UNDP, Reducing Disaster Risk: A Challenge for Development—A Global Report. United Nations Development Programme, 164 pages, 2004.
[66] USGS, http://earthquake.usgs.gov/eqcenter/pager/. Downloaded, 30 May 2007.
[67] L.J. Vale, T.J. Campanella, M.W. Fishwick, The resilient city: how modern cities recover from disaster. The Journal of American Culture, 28(4):456, 2005.
[68] G.M. Weiss, Mining with rarity: a unifying framework. SIGKDD Explorations, 6(1):7–19, 2004.
[69] G.M. Weiss, H. Hirsh, Learning to predict rare events in event sequences. In: Proceedings of the Fourth International Conference on Knowledge Discovery and Data Mining, pp. 359–363, 1998.

[70] R. White, D. Etkin, Climate change, extreme events and Canadian insurance. Natural Hazards, 16(2–3):135–163, 2004.
[71] F.W. Zwiers, H. von Storch, On the role of statistics in climate research. International Journal of Climatology, 24(6):665–680, 2004.

Chapter 14
TinyOS Education with LEGO MINDSTORMS NXT

Rasmus Ulslev Pedersen

Abstract The LEGO MINDSTORMS NXT[1]—*armed* with its embedded ARM7 and ATmega48 microcontrollers (MCUs), Bluetooth radio, four input ports, three output ports, and dozens of sensors—is proposed as an educational platform for TinyOS.[2] The purpose of this chapter is to assess NXT for use in wireless sensor network education. To this end, the following items are evaluated: NXT hardware/software, LEGO MINDSTORMS "ecosystem", and educational elements. We outline how this platform can be used for educational purposes due to the wide selection of available and affordable sensors. For hardware developers, the ease of creating new sensors will be hard to resist. Also, in the context of education, TinyOS can be compared to other embedded operating systems based on the same hardware. This chapter argues that this comparability facilitate across-community adoption and awareness of TinyOS. Finally, we present the first TinyOS project on NXT, hosted both at TinyOS 2.x contrib and SourceForge under the *nxtmote* name.

14.1 Introduction

Established and emerging embedded operating systems can benefit from the implementation of LEGO MINDSTORMS NXT for several reasons. The open source policy of the recently released NXT makes it easy to port new operating systems to NXT. NXT can be seen in Figs. 14.1 and 14.6. The sensors can be seen in Figs. 14.2, 14.3, and 14.5.

The Mindstorms NXT system from LEGO is an interesting and flexible hardware platform. The NXT PCB is equipped with an ARM MCU, as well as an AVR MCU; both MCUs are from Atmel. Furthermore, these two MCUs are connected to input

R.U. Pedersen
Department of Informatics, Copenhagen Business School, Copenhagen, Denmark
e-mail: rup.inf@cbs.dk

[1] http://mindstorms.lego.com/.

[2] http://www.tinyos.net.

ports, output ports, an LCD, Bluetooth radio, and USB. There is already a rich set of sensors available from both LEGO and third-party vendors which enable the use of the NXT system for prototyping almost any conceivable education (or research) project.

The NXT brick and LEGO Mindstorms have a large user base. The NXT brick provides open source firmware code written in the programming language C for both HCUs on the board. It is given that the majority of different embedded operating systems—besides TinyOS—will be ported to the NXT brick. This is valuable as it could possibly become the first platform where TinyOS can be directly compared to many other embedded operating systems. An idea of porting TinyOS to the older RCX platform can most likely be attributed back to a project idea list in a class taught by David Culler in 2003.[3] NXT is better than RCX in many ways. LEGO's open source policy is one new enabling factor.

This chapter recommends NXT as an educational platform. Therefore, we present a technical aspect of NXT in each section and then discuss why this is interesting from a wireless sensor networks educational perspective.

The structure of the chapter is such that we start with a an overview of the NXT hardware, software and sensors in Sect. 14.2. The educational perspectives are outlined in each subsection. This is followed by Sect. 14.3, which is a description of how NXT covers several operating systems, with more to come. The open source development for NXT is presented along with some machine learning aspects for this system. Finally, we outline a TinyOS platform for NXT in Sect. 14.4.

14.2 NXT Brick

This section briefly describes the NXT software, hardware and sensor architecture. LEGO is publishing a detailed set of documentation that is available from the Mindstorms web site.[4] We also cover a few of the firmware operating system replacements and some of the available front-ends that are already emerging for NXT.

The NXT PCB in Fig. 14.1 highlights some of the main components: (a) ARM7 MCU, (b) ATmega48 MCU, (c) CSR BlueCore4 Bluetooth radio, (d) SPI bus and touchpad signals, (e) high-speed UART behind input port 4 w. I2C/TWI, (f) output (generally motor) port, (g) USB port, (h) four-button touchpad, and (i) 100x64 LCD display.

14.2.1 LEGO NXT Software Architecture

The standard firmware NXT virtual machine (VM) executes programs consisting of NXT bytecodes. These bytecodes are grouped into six classes: *math, logic, comparison, data manipulation, control flow, and system i/o*. The NXT VM executes

[3] http://www.cs.berkeley.edu/∼culler/cs252-s03/projects.html.

[4] http://mindstorms.lego.com/.

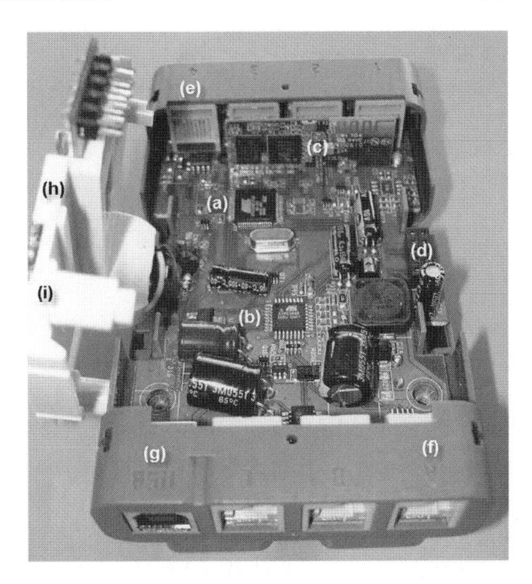

Fig. 14.1 NXT with important HW highlighted

as many bytecodes as it can in an update round in the NXT OS. During this update round each module can perform its functions, such as updating the display in the case of the display module. The approach taken with the NXT software is to address shared structures and subsequently each module is responsible for its own actions. This round-robin approach dictates a need for explicit state-control in many of the modules.

Even though we promote TinyOS as the operating system for educational wireless sensor networks, it is valuable to have the LEGO operating system available for comparison purposes. The learning curve is very flat if students start out with the drag-and-drop based NXT-G block programming language (see Fig. 14.2). After this, they progress to playing with and understanding the firmware sources, which are written in ANSI C. Finally, they can move on to TinyOS by simply flashing NXT with a new firmware image. Having one hardware platform makes it much easier for the student to see how the software differs. One example of an interesting exercise would be to compare Maté [2] with the LEGO MINDSTORMS NXT virtual machine (as defined in c_cmd.c of ARM7 NXT source files).

14.2.2 NXT Hardware Architecture

NXT features an ARM7 processor and an AVR: the AT91SAM7S256 and the ATmega48 as listed in Table 1. It also has a 100×64 LCD, and a Bluetooth (BT) radio that share the same SPI bus. The Bluetooth chip is from CSR—named BlueCore—and it is running the BT-stack from CSR programmed with BlueLab. The AVR controls the three output ports and communicates with the ARM over I^2C/TWI.

Table 1 NXT MCUs: ARM7 and AVR

	AT91SAM7S256	ATmega48
Flash	256 KiB	4 KiB
SRAM	64 KiB	512 B
Speed	48 MHz	8 MHz

As should be clear from the hardware presentation, this platform is rich enough to cover the needs of most university wireless sensor networks classes. The powerful ARM7 is the workhorse of the system, and it is the one on which students will have the most practice learning how a TinyOS platform is composed from a software point of view. Advanced students can also flash the ATmega48 and make this co-processor do new things. If NXT can host TinyOS implementations on both MCUs, then we can even have a "dual-core" TOS implementation. This constellation can be used to determine optimal power-consumption depending on the state of the system during some deployments. Some of the challenges regarding when to send and when to listen on the radio would be the same for the I2C bus between these two MCUs. Problems like these are well-suited for advanced student exercises.

14.2.3 Sensors

The input and output ports feature a six-wire RJ12 connector. On the input ports there are both analog and digital lines. On the output ports, PWM functionality is used with the motors in the standard NXT firmware. The NXT comes with a basic set of sensors. This basic set includes an ultrasonic distance measurement sensor, a light intensity sensor, a sound sensor, a touch sensor, and motors. Furthermore, there are a number of third-party sensors available, such as various acceleration sensors, a compass sensor, a temperature sensor, etc. More complete listings are available at the Mindsensor and Hitecnic web sites. The US-based Vernier also produces sensors such as pH probes or magnetic field probes to name just a few.

With NXT comes a set of standard sensors, as can be seen in Fig. 14.2. This set of sensors can give most students enough to work with to create their custom motes: the light and microphone sensor are almost standard for a mote. It should not go unnoticed that there are three motors (see Fig. 14.2b), which make it simple to create dynamic moving and data-collecting motes. For completeness, there is one NXT-G block shown, which is used in the block-based programming language.

Mindsensors is one of the main providers of sensors for NXT. Figure 14.3 shows some additional sensors that are available for NXT. The motes, which can be built into the WSN classes with these sensors, resemble standard motes like MICA because we now have a temperature sensor. With the magnetic compass sensor it is possible to build a dynamic mote for outdoor navigation/orientation. Finally, with the sensor building kit in Fig. 14.3e, there is actually a way to build a new sensor in

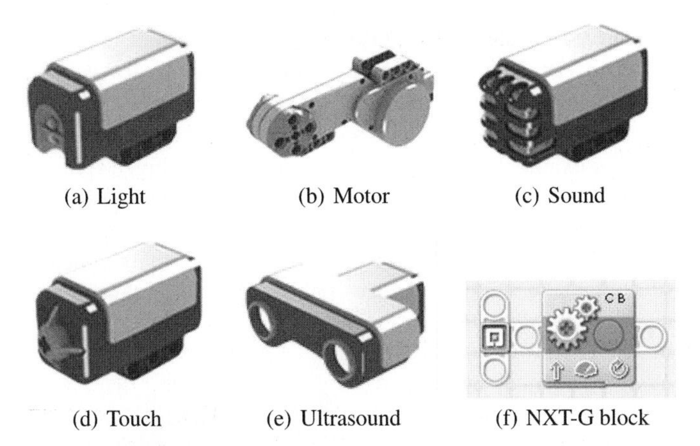

(a) Light (b) Motor (c) Sound

(d) Touch (e) Ultrasound (f) NXT-G block

Fig. 14.2 Standard LEGO MINDSTORMS NXT sensors and NXT-G block

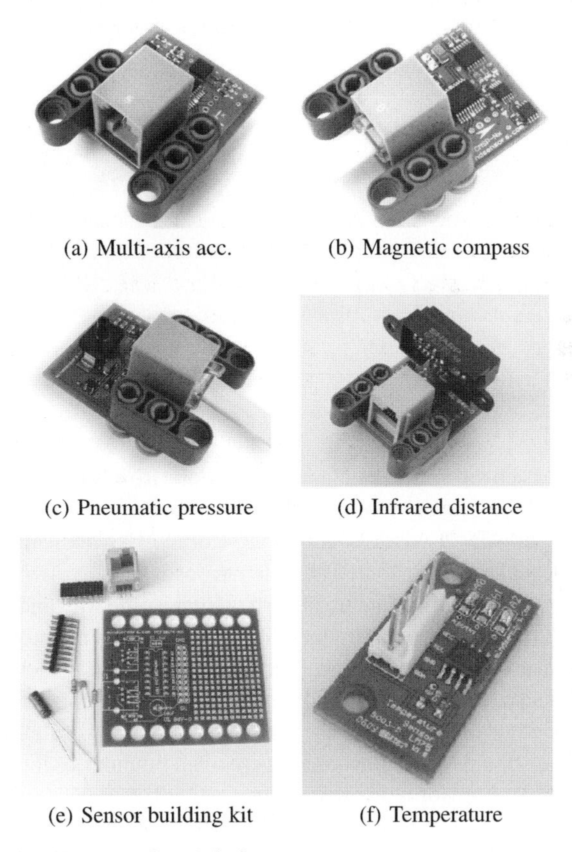

(a) Multi-axis acc. (b) Magnetic compass

(c) Pneumatic pressure (d) Infrared distance

(e) Sensor building kit (f) Temperature

Fig. 14.3 Selected NXT sensors from Mindsensors

an affordable and accessible manner. One of the advantages of multiple communities "sharing" NXT HW is that mass-production drives prices down to a level where classroom teaching becomes affordable.

A second provider named Hitechnic supplements and adds further possibilities for education. For example, with their prototype board in Fig. 14.4d, we can now let students experiment with I2C-based sensors. If a logic analyzer is available, then this could also be used to demonstrate how I2C and SPI look at the digital level. NXT only has a limited number of ports, but with the multiplexer there is a way to achieve the "sandwich" of multiple sensors on one mote. With this multiplexer there are many opportunities to try different setups.

(a) Color (b) Gyro

(c) Multiplexer (d) Prototyping board

Fig. 14.4 Sensors from Hitechnic

For an applied sensor networks class, the sensors from Vernier come in handy. One advantage of the sensors from vendors like Vernier (see Fig. 14.5) is that it may be possible to offer a wireless sensor networks class dedicated to creating a real-world functional mote for some valuable purpose (such as environmental monitoring). Three of the above sensors involve oxygen in some way and can be used for different indoor and outdoor experiments. Soil moisture and magnetic field sensors also have their uses. These sensors are only examples, and at the time of writing the Vernier NXT converter is in its final testing stage. They are included here only to indicate the idea of a nearly unlimited number of sensors available from Venier and others, which should break down most perceived boundaries in terms of what is possible with wireless sensor networks.

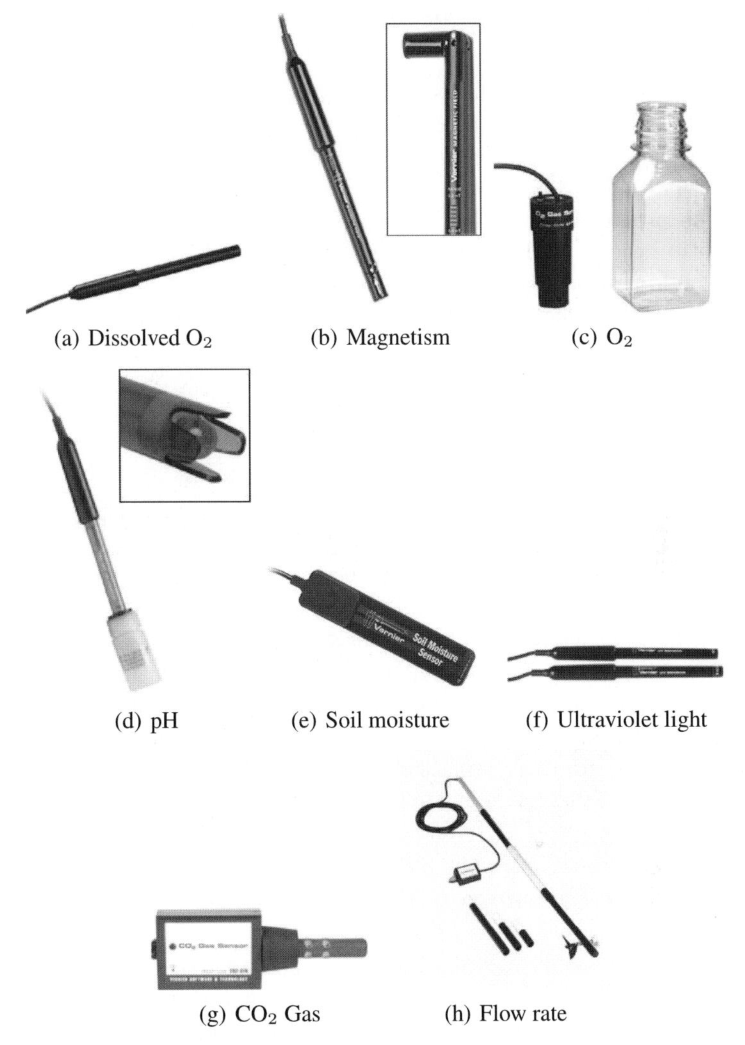

(a) Dissolved O_2 (b) Magnetism (c) O_2

(d) pH (e) Soil moisture (f) Ultraviolet light

(g) CO_2 Gas (h) Flow rate

Fig. 14.5 Various Vernier sensors

14.3 LEGO Ecosystem

LEGO offers a vibrant NXT community. There are multiple blog and forum sites. Many individuals and groups work on different firmware replacements and other embedded operating systems. We believe a system like NXT motivates students to work and thus makes the student invest more energy in learning. This can be beneficial for reaching higher levels of understanding as framed in Bloom's taxonomy.[5]

[5] http://en.wikipedia.org/wiki/Bloom's_Taxonomy.

14.3.1 NXT Firmware Replacements and Front-ends

A non-exhaustive list here includes a modified version of the firmware, which can be programmed using RobotC. Front-ends to the standard firmware are also becoming available, and in this category we find the NXT Bytecodes, NBC, which is available from the bricxx Sourceforge project. The Microsoft Robotics Studio is also a choice for programming the NXT. It is evident that numerous firmware replacements are to become available featuring most embedded operating systems, and probably multiple front-ends using most programming languages. For students using TinyOS on NXT, there is potentially much value from looking into how others have done a certain hardware tweak.

14.3.2 Toolchains

The LEGO NXT firmware is released for an IAR systems toolchain compiler. We have converted the source code and made the gcc-based toolchain available,[6] as announced at the LEGO Users Group Network (LUGNET™). The toolchains are for the ARM7 (arm-elf-gcc) and later also for the AVR (avr-gcc).

Most of the existing motes have a gcc-based toolchain. However, it is also good for a student to see a compiler like IAR's (which LEGO uses) in action and compare the compiled code to gcc. It can be that gcc is not as good at creating a small image, which can give the student motivation to play with and tweak the gcc compiler flags. Many of the NXT firmware replacement projects use gcc, so there are different toolchains available that can be fun to look at as well.

14.3.3 Machine Learning on NXT

The TinyOS motes could (like so many other embedded systems such as embedded Java) be effective platforms for experimenting with energy-aware machine learning algorithms [3]. NXT is equipped with the powerful ARM MCU and the smaller AVR MCU. This setup makes it possible to create power-aware configurations seeking the optimal balance between using the somewhat restricted AVR versus the more powerful ARM, as new obstacles or decisions call for more computational capability during deployment of the mote. In addition, the ARM features two instruction sets. One is the 16-bit Thumb instruction set that has a smaller code footprint, while the 32-bit ARM© instruction set provides higher performance. The ARM© instruction set also delivers several DSP-like instructions that provide for low-level power and performance optimizations.

From an educational perspective, it makes sense to identify machine learning as an important field for TinyOS on NXT. The instruction set on the ARM is perfect for handcrafting an efficient assembler-based algorithm and comparing the power

[6] http://nxtgcc.sf.net.

consumption with what gcc compiles. This is not unique to NXT, of course, but if the AVR is brought into consideration, then it is possible to compare instruction sets and even to measure and balance power consumption. Should this not be enough for some students, then the prototyping boards (or a CAD program like the free EagleCAD) can be used to forge a dedicated number-crunching sensor using specialized hardware components.

14.3.4 Programming and Debugging

Both the ARM and AVR can be accessed by JTAG connectors. The J17 JTAG pins for the ARM are not mounted on the production PCB; however, it is possible to create one using a 1.27 mm pitch row connector. The square hole is pin 1 on the PCB. Each of the eight pins from J17 can be connected to a standard 20 port JTAG like this: $1 \rightarrow 9$, $2 \rightarrow 7$, $3 \rightarrow 13$, $4 \rightarrow 15$, $5 \rightarrow 5$, $6 \rightarrow 4$, $7 \rightarrow$ not connected, and $8 \rightarrow 1$. This works even without soldering the pins to J17. Subsequently, several options exist for flashing the ARM7: J-Link from IAR or Segger, ARM-JTAG from Olimex, or the JTAGkey from Amontec to name a few.

The NXT ARM is easily flash programmed with the standard USB cable and the LabVIEW-based block programming IDE from LEGO. A more direct approach is to use the SAM-BA program from Atmel or a open source called LibNXT (a project in Google Code) that includes flash/RAM programming support for both Linux and Windows. Students will perhaps find debugging with JTAG a little difficult on NXT. Still it would be possible to debug the ARM7 over the serial line using a GDB stub. The AVR can be programmed using a suitable system like the AVR ISP In-System Programmer from Atmel.

14.3.5 Bluetooth Radio

The Bluetooth radio is a CSR BlueCore4 chip. It supports scatternets and piconets. The LEGO NXT uses a limited subset of the BlueCore functionality. The BlueLab SDK can reprogram the virtual machine in the BC417-143BQN chip (see Fig. 14.1c). The virtual machine separates the code from the basic operation of the chip. Note that the virtual machine is more like a sandbox separating user code from firmware code than a virtual machine in the sense of a Java VM or Maté.

Bluetooth is an important technology and has been widely adopted outside wireless sensor networks. From an educational standpoint, it is not important if NXT had Bluetooth, Zigbee, WiFi or whatever. What is important is that it has a radio mounted on the standard PCB and it is easy to connect other radios to the sensor ports. This is a prerequisite for using NXT as a wireless sensor network platform. It is not necessary to have many NXTs connect to form a wireless sensor network for a class of students. However, there will be situations where it is better to have many NXTs connect, like a sensor network composed of many motes.

14.3.6 Additional Radio Sensor(s)

There are several approaches to adding radio sensors. It is possible to tap into the SPI bus that is shared between the LCD and the BlueCore modules; this requires some additional control lines, probably coming from two of the input ports (each port offers two GPIO lines). A better option is to use one of the input ports to create an I^2C based sensor with an additional MCU involved.

Connecting a radio to NXT is something which could be a project following a advanced class on wireless sensor networks. There are development modules ready from vendors like Chipcon which could make this possible. The CC2420 is a candidate since it has a well-documented stack, and is compatible with many other motes, not to mention the CC2430 with built-in MCU support.

14.4 Proposing a TinyOS Educational Platform for NXT

The proposed platform is labeled *nxtmote*.[7] One of the good things about the TinyOS community is the intent to share code. Therefore, most of the pieces needed to assemble a new mote for NXT are already available. The most important parts for NXT are the ARM MCU and the Bluetooth radio. After these come additional components like an 802.4.15 radio sensor.

The main MCU, which is the ARM7, can be based on imote2. There are pros and cons to this, but from an educational standpoint it could make sense to have two similar implementations. The Bluetooth Radio implementation can to some extent mimic the TinyBT [1] stack, even though there is a difference as the Bluetooth radio virtual machine is pre-programmed by LEGO to work in a certain way.

A second generation nxtmote can include a second TinyOS core running on the ATmega48. It can also include a Zigbee radio or another 802.4.15 radio attached to one of the ports. In addition, the BlueCore chip could perhaps be reprogrammed to expose a wider part of the functionality than is presently available in the standard firmware.

The nxtmote project was from the beginning intended as an educational project, which is why it has been defined within the TinyOS contrib CVS.[8] tree. Both the source code and the nxtmote tutorials are hosted here.

14.5 Conclusion

The NXT brick is a suitable platform for prototyping different TinyOS experiments for a wide variety of sensors. Figure 14.6 shows the NXT with the standard sen-

[7] Introduced at the EWSN 2007 poster session. The project home page url is: http://nxtmote.sourceforge.net.

[8] http://tinyos.cvs.sourceforge.net/*checkout*/tinyos/tinyos-2.x-contrib/index.html/.

Fig. 14.6 NXT with standard LEGO sensors

sors attached. We have presented the first actual TinyOS project for this research direction some years after it was first formulated (see the Introduction). The timing is now more optimal as NXT is very different from the older RCX both in terms of hardware and the open source policy laid forth by LEGO. With nxtmote it will be possible to add a dynamic (i.e. robotic) element to the experiments as well as a machine learning component. Finally, the great availability of NXT and compatible sensors ensures an affordable educational platform for most embedded operating systems including TinyOS.

References

[1] M. Leopold, Tiny bluetooth stack for TinyOS. Technical report, Department of Computer Science, University of Copenhagen, 2003.

[2] P. Levis, D. Culler, Maté: a tiny virtual machine for sensor networks. SIGOPS Operating Systens Review, 36(5):85–95, 2002.

[3] R. Pedersen, Distributed support vector machine. PhD thesis, Department of Computer Science, University of Copenhagen, 2005.

Index

Printing: Krips bv, Meppel
Binding: Stürtz, Würzburg